# HODDER
# *mathematics*
## INTERMEDIATE
## SECOND EDITION

## 2

**Series editor: Roger Porkess**

**Catherine Berry**
**Pat Bryden**
**Diana Cowey**
**Dave Faulkner**
**Julian Thomas**
**Christine Wood**

MEI

Hodder & Stoughton
A MEMBER OF THE HODDER HEADLINE GROUP

## *Acknowledgements*

The authors and publishers would like to thank the following companies, agencies and individuals who have given permission to reproduce copyright material: NEAB (AQA). Every effort has been made to trace and acknowledge ownership of copyright. The publishers will be glad to make suitable arrangements with any copyright holder whom it has not been possible to contact.

Illustrations were drawn by Josephine Blake, Maggie Brand, Tom Cross, Bill Donohoe, Jeff Edwards, Joe McEwan and Mark Walker of Ian Foulis and Associates.

Photos supplied by the Health Education Authority (page 92), Life File (page 176 top), Jon Woodhouse/ Life File (page 176 bottom), Andrew Ward/Life File (page 179), Emma Lee/Life File (page 180).

Page design and cover design by Lynda King.

Orders: please contact Bookpoint Ltd, 130 Milton Park, Abingdon, Oxon OX 14 4SB.
Telephone: (44) 01235 827720, Fax: (44) 01235 400454. Lines are open from 9.00 – 6.00, Monday to Saturday, with a 24 hour message answering service.
You can order through our website www.hodderheadline.co.uk

*British Library Cataloguing in Publication Data*

A catalogue record of this title is available from The British Library

ISBN 0 340 803746

First published 1998

Second edition 2002

| Impression number | 10 9 8 7 6 5 4 3 2 |
|---|---|
| Year | 2008 2007 2006 2005 2004 2003 |

Copyright ©1998, 2002 Catherine Berry, Pat Bryden, Diana Cowey, Dave Faulkner, Julian Thomas and Christine Wood.

All rights reserved. No part of this publication may be reproduced or transmitted in any form or by any means, electronic or mechanical, including photocopy, recording, or any information storage and retrieval system, without permission in writing from the publisher or under licence from the Copyright Licensing Agency Limited. Further details of such licences (for reprographic reproduction) may be obtained from the Copyright Licensing Agency Limited, of 90 Tottenham Court Road, London W1T 4LP.

Cover photo from Photonica.

Typeset by Pantek Arts Ltd, Maidstone, Kent.

Printed in Italy for Hodder & Stoughton Educational, a division of Hodder Headline, 338 Euston Road, London NW1 3BH.

# Introduction

This is the second of two textbooks covering Intermediate Tier GCSE. Students following a two year course would expect to take one year on each book. The books cover the requirements of Intermediate Tier GCSE and so are suitable for use with any specification. The division of material between them ensures both coverage of the modules within the MEI GCSE specification and a balanced teaching curriculum over the two years. This book also covers the mathematics requirements of GNVQ Application of Number at Level 3.

This is the second edition of this book. It has been adapted to take account of new GCSE criteria. These apply to courses with first teaching in September 2001 and first certification in summer 2003. The changes are considerable and this new edition incorporates all of them.

The book is divided into 16 chapters, forming a logical progression through the material (some teachers may however wish to vary this order). Each chapter is divided into a number of double-page spreads, designed to be teaching units. The material to be taught is covered on the left-hand page; the right-hand pages consist entirely of work for the students to do. Each chapter ends with a mixed exercise covering all of its content. Further exercise sheets and tests are provided in the Teacher's Resource.

The instruction (i.e. left-hand) pages have been designed to help teachers engage their students in whole class discussion. The symbol ⁇ is used to indicate a Discussion Point; teachers should see it as an invitation.

Most of the right-hand pages end with a practical activity. These are suitable for both GCSE and GNVQ students; some can be used for portfolio tasks. Advice on these is available in the Teacher's Resource and, where relevant, raw data is also supplied. Most students will not do all of the activities (they can be quite time-consuming) but the authors think it is important that they do as many of them as possible; they connect the mathematics classroom to the outside world and to other subjects.

Where knowledge is assumed, this is stated at the start of the chapter. There is a general expectation that students will know the content of Foundation Tier GCSE. Questions indicated with a calculator icon 🖩 need to be answered with a calculator. The 'no calculator' icon ⌧ indicates that a calculator should definitely not be used. Many questions have neither icon and these require a sensible judgement. Students should do as many of these as possible without a calculator in order to practise for the non-calculator GCSE questions. However, they also need to work through plenty of questions and using a calculator often allows them to work faster.

Although students are to be encouraged to use I.T., particularly spreadsheets, specific guidance is limited to the Teacher's Resource. Otherwise, the book would have been based on one particular package to the frustration of those using all the others.

The authors would like to thank all those who helped in preparing this book, particularly Chris Curtis for his advice on early versions of the manuscript, and Karen Eccles who typed much of the first edition.

# Contents

**Information pages** ............ 2

## Chapter One: Fractions, decimals and percentages ............ 4
*Reminder* ............ 4
*Multiplying fractions* ............ 6
*Dividing fractions* ............ 8
*Approximate percentages* ............ 10
*Finding the original price* ............ 12
*Percentage problems* ............ 14
*Finishing off* ............ 16

## Chapter Two: Formulae and equations ............ 20
*Reminder* ............ 20
*Making up formulae* ............ 22
*Working with unknowns* ............ 24
*Using equations to solve problems* ............ 26
*Using graphs to solve equations* ............ 28
*Trial and improvement* ............ 30
*Rearranging a formula* ............ 32
*Finishing off* ............ 34

## Chapter Three: Triangles and polygons ............ 36
*Reminder* ............ 36
*Angles and triangles* ............ 38
*Drawing triangles* ............ 40
*Drawing more triangles* ............ 42
*Polygons* ............ 44
*Angle sum of a polygon* ............ 46
*Finishing off* ............ 48

## Chapter Four: Graphs and co-ordinates ............ 50
*Working with co-ordinates* ............ 50
*Using co-ordinates* ............ 52
*Graphs for real life* ............ 54
*Number patterns and sequences* ............ 56
*Finishing off* ............ 58

## Chapter Five: Simultaneous equations ............ 60
*Using simultaneous equations* ............ 60
*More simultaneous equations* ............ 62
*Multiplying both equations* ............ 64
*Other methods of solution* ............ 66
*Finishing off* ............ 68

## Chapter Six: Trigonometry ............ 70
*Introduction to trigonometry* ............ 70
*Using tangent (tan)* ............ 72
*Finding the adjacent side using tan* ............ 74
*Using sine (sin)* ............ 76
*Using cosine (cos)* ............ 78
*Using sin, cos and tan* ............ 80
*Using trigonometry* ............ 82
*Finishing off* ............ 84

## Chapter Seven: Inequalities ............ 86
*Using inequalities* ............ 86
*Number lines* ............ 88
*Solving inequalities* ............ 90
*Inequalities and graphs* ............ 92
*Regions bounded by sloping lines* ............ 94
*Solution sets* ............ 96
*Finishing off* ............ 98

## Chapter Eight: Indices and standard form ............ 100
*Reminder* ............ 100
*Rules of indices* ............ 102
*Calculations using standard form* ............ 104
*Finishing off* ............ 106

# Contents

**Chapter Nine: Circles and tangents** .................. 108
- Shapes in a circle .................. 108
- Angles in a circle .................. 110
- Tangents .................. 112
- Finishing off .................. 114

**Chapter Ten: Manipulating expressions** .................. 116
- Like terms .................. 116
- Factorising .................. 118
- Expanding two brackets .................. 120
- Squares .................. 122
- Finishing off .................. 124

**Chapter Eleven: Probability** .................. 126
- Reminder .................. 126
- Two outcomes: 'either, or' .................. 128
- Two outcomes: 'first, then' .................. 130
- Probability trees .................. 132
- Finishing off .................. 134

**Chapter Twelve: Locus** .................. 136
- Simple loci .................. 136
- A point equidistant from two fixed points .................. 138
- A point equidistant from two lines .................. 140
- Finishing off .................. 142

**Chapter Thirteen: Quadratics** .................. 144
- Factorising quadratic expressions .................. 144
- More quadratic factorisation .................. 146
- Quadratic equations .................. 148
- Finishing off .................. 150

**Chapter Fourteen: Transformations** .................. 152
- Reminder .................. 152
- Translations using column vectors .................. 154
- Reflection .................. 156
- Rotation .................. 158
- Combining transformations .................. 160
- Finishing off .................. 162

**Chapter Fifteen: Fractions in algebra** .................. 164
- Reminder .................. 164
- Indices .................. 166
- Rational functions .................. 168
- Finishing off .................. 170

**Chapter Sixteen: Enlargement and similarity** .................. 172
- Reminder .................. 172
- Centres of enlargement .................. 174
- Scale factors less than 1 .................. 176
- Similar shapes .................. 178
- Using similarity .................. 180
- Finishing off .................. 182

**Answers** .................. 184

# Information

## How to use this book

This symbol next to a question means you need to use your calculator.

This symbol next to a question means you are not allowed to use your calculator.

This symbol means you will need to think carefully about a point and may want to discuss it.

### Triangles

An **equilateral** triangle has 3 equal sides.

An **isosceles** triangle has 2 equal sides.

A **scalene** triangle has no equal sides.

A **right-angled** triangle has 1 right angle.

An **acuted-angled** triangle has 3 acute angles.

An **obtuse-angled** triangle has 1 obtuse angle.

$$\text{Area of a triangle} = \frac{1}{2} \times \text{base} \times \text{height}$$

### Quadrilaterals

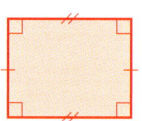

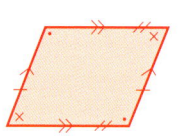

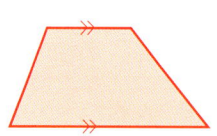

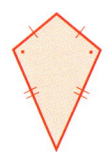

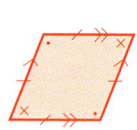

square   rectangle   parallelogram   trapezium   kite   rhombus

Area of a parallelogram = base × vertical height

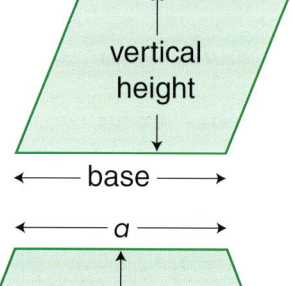

Area of a trapezium = $\frac{1}{2}(a + b)h$

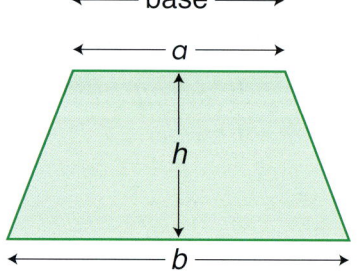

# Circles

Circumference of circle = π × diameter
= 2 × π × radius

Area of circle = π × (radius)$^2$

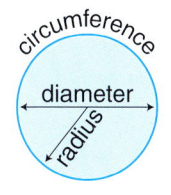

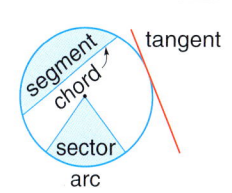

# Solid figures

**Volume of cuboid = length × width × height**

**Volume of prism = area of cross section × length**

**Volume of cylinder = $\pi r^2$ × length**

# Trigonometry

$\sin \theta = \dfrac{\text{opposite}}{\text{hypotenuse}}$

$\cos \theta = \dfrac{\text{adjacent}}{\text{hypotenuse}}$

$\tan \theta = \dfrac{\text{opposite}}{\text{adjacent}}$

Pythagoras' rule: $x^2 + y^2 = h^2$

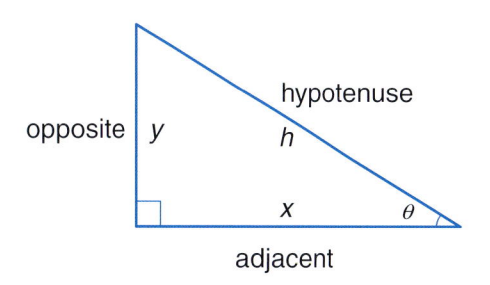

# Units

## Metric system

### Length

k   1 kilometre = $10^3$ metres = 1000 metres

h   1 hectometre = $10^2$ metres = 100 metres

da   1 decametre = $10^1$ metres = 10 metres

d   1 decimetre = $10^{-1}$ metres = $\dfrac{1}{10}$ metre

c   1 centimetre = $10^{-2}$ metres = $\dfrac{1}{100}$ metre : 100 centimetres = 1 metre

m   1 millimetre = $10^{-3}$ metres = $\dfrac{1}{1000}$ metre : 1000 millimetres = 1 metre

The units for mass and capacity follow the same pattern. Notice that:
1 kilogram = 1000 grams          1 litre = 1000 millilitres

Notice also that: 1 tonne = 1000 kg

## Imperial

12 inches = 1 foot          16 ounces = 1 pound
3 feet = 1 yard          14 pounds = 1 stone
1760 yards = 1 mile          8 stones = 1 hundredweight (cwt)
                         20 cwt = 1 ton

# One

# Fractions, decimals and percentages

**Before you start this chapter you should be able to**

- ★ convert between improper fractions and mixed numbers
- ★ add and subtract fractions
- ★ do calculations involving decimals
- ★ find a fraction of a quantity
- ★ find a percentage of a quantity
- ★ compare using fractions, decimals or percentages.

Use the following questions to check that you still remember these topics.

## Reminder

- $\frac{21}{8}$ is an **improper fraction** (or top heavy fraction)
- $2\frac{5}{8}$ is a **mixed number**

**Revision exercise**

**1** A sports hall is open 16 hours a day. This pie chart shows today's bookings. What fraction of time is booked to each activity?

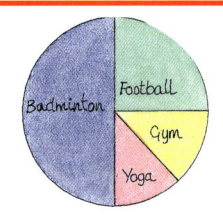

**2** Change these improper fractions to mixed numbers.

a) $\frac{11}{4}$   b) $\frac{13}{6}$   c) $\frac{29}{8}$

d) $\frac{14}{3}$   e) $\frac{22}{7}$   f) $\frac{18}{5}$

**3** Changes these mixed numbers into improper fractions.

a) $3\frac{1}{4}$   b) $3\frac{7}{8}$   c) $2\frac{7}{10}$

d) $5\frac{2}{3}$   e) $7\frac{2}{5}$   f) $4\frac{13}{16}$

**4** Work out

a) $2\frac{7}{8} + \frac{5}{8}$   b) $1\frac{3}{4} + 3\frac{3}{8}$   c) $3\frac{7}{16} - 2\frac{1}{8}$   d) $5\frac{1}{4} - 3\frac{11}{16}$

e) $2 - \frac{13}{16}$   f) $4\frac{5}{8} + 1\frac{1}{2}$   g) $4\frac{3}{4} - 1\frac{7}{8}$   h) $1\frac{1}{2} + 2\frac{2}{3}$

i) $4\frac{7}{12} - 2\frac{1}{4}$   j) $4\frac{2}{3} + 2\frac{5}{6}$   k) $3\frac{7}{10} - 2\frac{1}{5}$   l) $5\frac{3}{5} + 2\frac{1}{4}$

## 1: Fractions, decimals and percentages

**5** Arrange these numbers in order of size starting with the smallest.

a) $\frac{11}{20}, \frac{51}{100}, \frac{27}{50}, \frac{13}{25}$

b) $\sqrt{26}, 5.1, 5\frac{1}{5}, \frac{41}{8}$

**6** Jodie is doing a survey for a video shop. She asks 80 people about the videos they like to watch. Three quarters like Horror films.

a) How many people is this?

b) Three quarters of the people who like Horror films watch more than one a month. How many people is this?

c) Two thirds of the people who like Horror films also like Action films. How many people is this?

**7** Work out

a) $3.4 + 1.72$  b) $6.59 - 2.4$  c) $0.2 \times 7$  d) $4.8 \times 5.1$

e) $2.3 \times 0.74$  f) $0.65 \times 10$  g) $725 \div 10$  h) $6 \div 0.2$

i) $3.6 \div 0.08$  j) $4.2 \times 0.25$  k) $240 \times 100$  l) $8410 \div 100$

m) $0.17 \div 0.6$  n) $0.3 \times 0.2$  o) $3.12 \times 0.05$  p) $9.07 \times 1000$

**8** Work out

a) 20% of 158   b) 7.5% of 600   c) 73% of £672.50

**9** The management and union are discussing a pay rise.

*Management offers 3.2%*

*UNION DEMANDS 4% Plus £20 per person*

Robert earns £250 a week and Emily earns £300 a week.

a) How much extra does Robert earn if the union gets what it demands?

b) What percentage increase is this?

c) What is Emily's new wage if the management offer is accepted?

d) What is Emily's new wage if the union gets what it demands?

**10** Philip and Samit are each doing a survey on smoking.
Philip finds 19 non-smokers out of 30 people questioned.
Samit finds 37 non-smokers out of 55 people questioned.
Compare their results.

*Revision exercise*

## 1: Fractions, decimals and percentages

# Multiplying fractions

Dave, Becky and Ravi share a pizza.

**?** *Dave has a quarter of it. How much is left?*

Becky and Ravi share the other three quarters.

Each has half of it.

Each gets $\frac{1}{2}$ of $\frac{3}{4}$:

$$\frac{1}{2} \times \frac{3}{4} = \frac{3}{8}$$

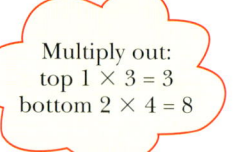

*Multiply out:*
*top 1 × 3 = 3*
*bottom 2 × 4 = 8*

**?** *Dave only eats half of his piece. What is $\frac{1}{2} \times \frac{1}{4}$?*

The pizza costs £8.00, but Ravi has a voucher for 30% discount.

Dave and Becky each work out how much this is.

Dave multiplies first and then divides top and bottom by 100.

$$30\% \text{ of } 800$$
$$= \frac{30}{100} \times \frac{800}{1}$$
$$= \frac{24000}{100}$$
$$= 240$$

$$30\% \text{ of } 800$$
$$= \frac{800}{100}$$
$$= \frac{240}{1} = 240$$

Becky divides top and bottom by 100 before multiplying. This is called **cancelling**.

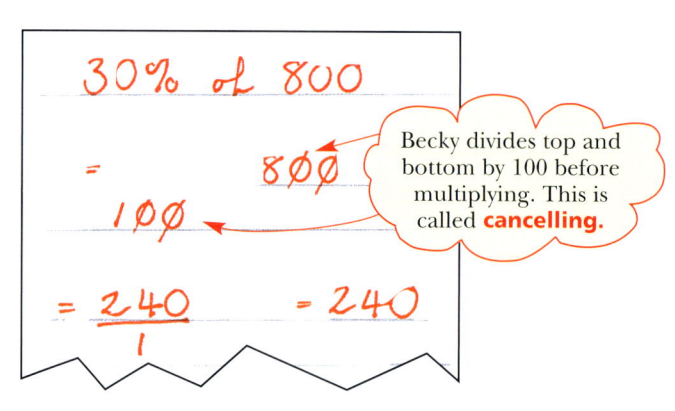

**?** *How much does Ravi pay for the pizza?*

The next two examples show you how to multiply mixed numbers.

Change $4\frac{1}{2}$ into an improper fraction

$$\frac{2}{3} \times 4\frac{1}{2}$$

Cancel

$$= \frac{\cancel{2}^1}{\cancel{3}_1} \times \frac{\cancel{9}^3}{\cancel{2}_1}$$

$$= \frac{3}{1} = 3$$

Multiply out and change back into a mixed number

$$2\frac{1}{3} \times 3\frac{3}{4}$$

Change the mixed numbers into improper fractions

$$= \frac{7}{\cancel{3}_1} \times \frac{\cancel{15}^5}{4}$$

Cancel

$$= \frac{35}{4}$$

$$= 8\frac{3}{4}$$

Multiply out and change back to a mixed number

## 1: Fractions, decimals and percentages

**1** Work out

a) $\frac{1}{2} \times \frac{1}{3}$    b) $\frac{1}{2} \times \frac{3}{8}$    c) $\frac{1}{4} \times \frac{3}{5}$    d) $\frac{3}{4} \times \frac{5}{6}$

e) $\frac{3}{8} \times \frac{2}{3}$    f) $\frac{6}{7} \times \frac{7}{10}$    g) $\frac{3}{8} \times \frac{5}{8}$    h) $\frac{3}{4} \times \frac{10}{1}$

**2** Work out

a) $\frac{1}{2}$ of 7    b) $\frac{3}{4}$ of 6    c) $\frac{1}{3}$ of 8    d) $\frac{2}{5}$ of 4

e) $\frac{5}{8}$ of 20    f) $\frac{2}{3}$ of 14    g) $\frac{3}{8}$ of 10    h) $\frac{5}{6}$ of 9

**3** Work out

a) $\frac{1}{2} \times 6\frac{1}{2}$    b) $\frac{3}{4} \times 4\frac{1}{2}$    c) $\frac{1}{3} \times 2\frac{5}{8}$    d) $2\frac{1}{2} \times \frac{7}{10}$

e) $2\frac{4}{5} \times \frac{5}{8}$    f) $\frac{3}{4} \times 5\frac{1}{3}$    g) $1\frac{1}{2} \times 2\frac{1}{2}$    h) $2\frac{1}{4} \times 3\frac{1}{2}$

i) $3\frac{2}{3} \times 1\frac{1}{2}$    j) $5\frac{1}{3} \times 3\frac{3}{4}$    k) $1\frac{3}{8} \times 3\frac{1}{2}$    l) $6\frac{2}{5} \times 1\frac{7}{8}$

**4** Amanda lives $2\frac{3}{4}$ miles from work. She works 5 days a week. How many miles does she cover, travelling to and from work, in a week?

**5** (i) circle with radius 14 cm    (ii) circle with radius 28 cm

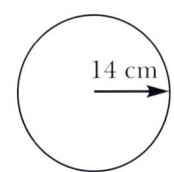

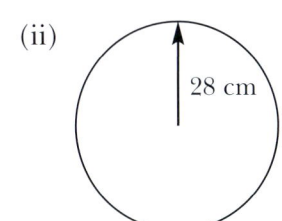

a) Using π = 22/7 find the circumference of each circle.

b) The radius of the second circle is twice the radius of the first. What has happened to the circumference?

c) Using π = 22/7 find the area of each circle.

d) The radius of the second circle is twice the radius of the first. What has happened to the area?

**6** Paula buys 60 lbs of boiled sweets. She makes up twenty $\frac{1}{4}$ lb bags, fifteen $\frac{1}{2}$ lb bags and fifteen $\frac{3}{4}$ lb bags. How much has she left over?

## 1: Fractions, decimals and percentages

# Dividing fractions

Becky and Ravi share $\frac{3}{4}$ of a pizza.

$$\frac{1}{2} \text{ of } \frac{3}{4} = \frac{1}{2} \times \frac{3}{4} = \frac{3}{8}$$

They each have $\frac{3}{8}$ of the pizza.

Another way of working this out is to say $\frac{3}{4}$ of a pizza is divided between 2 people.

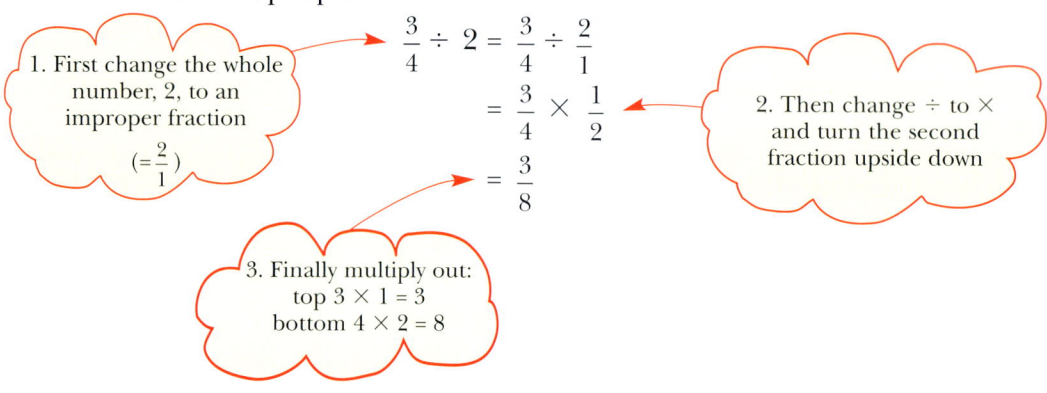

$$\frac{3}{4} \div 2 = \frac{3}{4} \div \frac{2}{1}$$
$$= \frac{3}{4} \times \frac{1}{2}$$
$$= \frac{3}{8}$$

1. First change the whole number, 2, to an improper fraction $(= \frac{2}{1})$

2. Then change ÷ to × and turn the second fraction upside down

3. Finally multiply out: top 3 × 1 = 3 bottom 4 × 2 = 8

Dave, Becky and Ravi share $4\frac{1}{2}$ chocolate bars equally.

How much does each person get?

Ravi works it out like this:

$4\frac{1}{2}$ is $\frac{9}{2}$ or 9 halves.
9 halves divided between 3 people gives 3 halves each.
3 halves is $\frac{3}{2}$ or $1\frac{1}{2}$. Each person has $1\frac{1}{2}$ chocolate bars.

Dave does it like this:

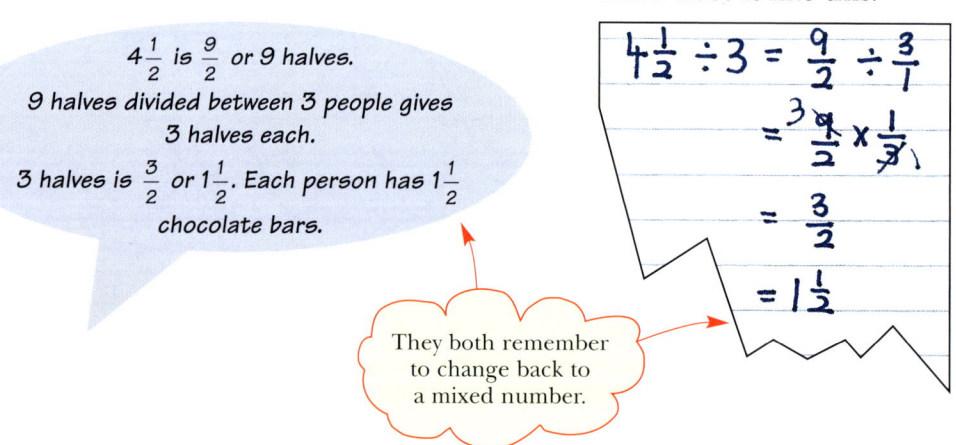

$$4\frac{1}{2} \div 3 = \frac{9}{2} \div \frac{3}{1}$$
$$= \frac{\cancel{9}^3}{2} \times \frac{1}{\cancel{3}_1}$$
$$= \frac{3}{2}$$
$$= 1\frac{1}{2}$$

They both remember to change back to a mixed number.

Here are two more examples of division.

$$5 \div \frac{3}{4} = \frac{5}{1} \div \frac{3}{4}$$
$$= \frac{5}{1} \times \frac{4}{3}$$
$$= \frac{20}{3} = 6\frac{2}{3}$$

$$4\frac{1}{2} \div 1\frac{1}{4} = \frac{9}{2} \div \frac{5}{4}$$
$$= \frac{9}{\cancel{2}_1} \times \frac{\cancel{4}^2}{5}$$
$$= \frac{18}{5} = 3\frac{3}{5}$$

## 1: Fractions, decimals and percentages

**1** Work out

a) $5 \div 4$  b) $\frac{1}{5} \div 2$  c) $1\frac{3}{5} \div 4$  d) $\frac{1}{3} \div 3$

e) $2\frac{1}{2} \div 5$  f) $2\frac{1}{4} \div 3$  g) $\frac{5}{8} \div 2$  h) $1\frac{1}{2} \div 6$

**2** Work out

a) $4 \div \frac{1}{3}$  b) $3 \div \frac{1}{2}$  c) $12 \div \frac{3}{4}$  d) $12 \div \frac{2}{5}$

e) $3\frac{1}{2} \div 4$  f) $2\frac{1}{4} \div 1\frac{1}{4}$  g) $3\frac{3}{4} \div \frac{3}{8}$  h) $2\frac{3}{16} \div 1\frac{1}{4}$

i) $2\frac{5}{8} \div 3\frac{1}{2}$  j) $8\frac{3}{4} \div 1\frac{1}{4}$  k) $6\frac{7}{8} \div 2\frac{3}{4}$  l) $12 \div 3\frac{1}{3}$

**3** A grocer buys pieces of cheese weighing 5 kg.

a) How many $\frac{1}{2}$ kg pieces can he get from this?

b) How many $\frac{1}{4}$ kg pieces can he get from it?

**4** A box is $12\frac{1}{2}$ inches long, 5 inches wide and $1\frac{1}{4}$ inches high.

Toy bricks are cubes with edges $1\frac{1}{4}$ inches long.

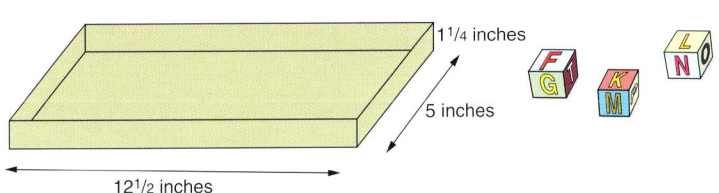

How many toy bricks can fit into the box?

**5** Jermaine's car does 35 miles per gallon and he has 6 gallons of petrol in the tank.

Jermaine's house — $3\frac{3}{4}$ miles — Fiona's house

How many times can he go to Fiona's house and back?

**6** A bookshelf is $29\frac{1}{4}$ inches long.

How many books can fit on the shelf if each book is

a) $\frac{3}{4}$ inch thick?  b) $1\frac{1}{8}$ inches thick?  c) $\frac{5}{8}$ inch thick?

## 1: Fractions, decimals and percentages

# Approximate percentages

In mathematics you will often need to calculate a percentage of a quantity.

When you do these calculations you should make rough calculations to estimate the size of the answer you expect.

Alex has this carpet in the middle of her lounge.

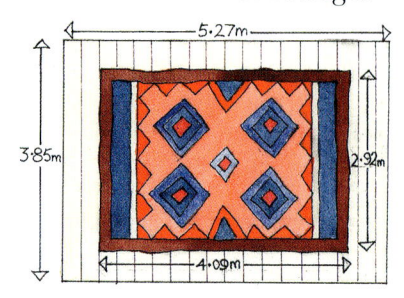

She estimates the percentage of floor left uncovered by the carpet.

First Alex estimates the area of the lounge.

Area of lounge (in m$^2$) = 5.27 × 3.85

*5.27 is about 5*    *3.85 is about 4*

≈ 5 × 4 = 20

Area of lounge is about 20 m$^2$.

Next Alex estimates the area of the carpet.

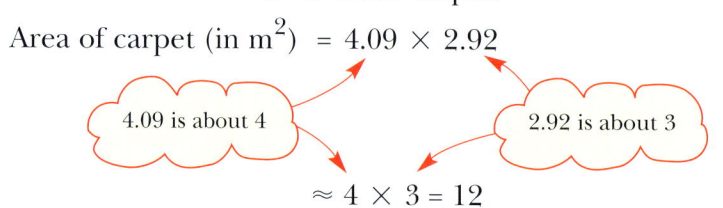

Area of carpet (in m$^2$) = 4.09 × 2.92

*4.09 is about 4*    *2.92 is about 3*

≈ 4 × 3 = 12

Area of carpet is about 12 m$^2$.

Now Alex can estimate the area of floor uncovered.

20 m$^2$ − 12 m$^2$ = 8 m$^2$

She uses these estimates to work out

percentage of floor uncovered = $\dfrac{\text{area of floor uncovered}}{\text{area of lounge}} \times 100$

$= \dfrac{8}{20} \times 100 = 40$

**About 40% of the floor is left uncovered.**

*Work out, correct to one decimal place, the percentage of the floor area which is left uncovered by the carpet. How good is the estimate?*

## 1: Fractions, decimals and percentages

**1** Ruth pays tax at 24p in the pound on her taxable income of £15 926.85.

a) Estimate how much tax she pays.
b) Work out, to the nearest penny, how much tax she pays.

Ruth's gross salary is £20 296.

c) Estimate the tax she pays as a percentage of her gross salary.
d) Work out, correct to one decimal place, the tax she pays as a percentage of her gross salary.

**2** A group of holidaymakers are asked which activity they enjoy most. Their responses are shown here.

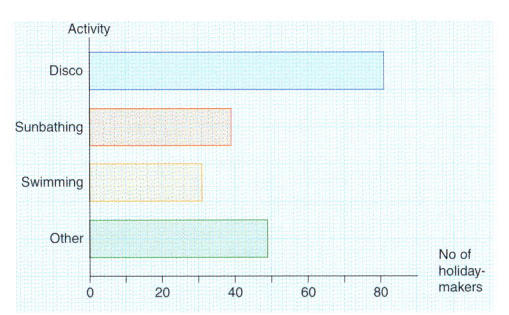

a) Estimate the number of responses.
b) What activity is chosen by about 40% of the respondents?
c) Estimate what percentage choose swimming.
d) Estimate what fraction of the holidaymakers choose 'other'.

**3** Ella has an allotment. Her old plot is 40 feet by 30 feet. She takes a new plot where the sides are 20% longer.

a) What are the dimensions of the new plot?
b) What is the area of the new plot?
c) Work out the increase in area as a percentage of the area of the old plot.
d) Her old rent, based on area, is £25.92 a year. Estimate her new rent.

**4** Charlotte's order is for £296.43.

a) Estimate how much discount she gets.
b) Work out exactly how much discount she gets.

Harry's order is for £1507.27.
He pays £1356.54.

c) Estimate how much discount he gets.
d) Estimate this discount as a percentage of the cost of the order.

☆☆BOOK WAREHOUSE☆☆
📖 CLOSING DOWN SALE 📖

orders over £200
6% discount
☺☺☺☺☺☺☺☺☺☺☺
orders over £1000
ask about our special discount!

## 1: Fractions, decimals and percentages

# Finding the original price

Peter has a clothes stall on the market.

He sells all of his stock at 40% profit.

*How much did he pay for this sweater?*

Peter has added 40% so

> 140% of the cost price is £49
>
> 1% of the cost price = $\frac{£49}{140}$ = £0.35
>
> 100% of the cost price = £0.35 × 100 = £35

**Peter bought the sweater for £35.**

'Jasmine's Jacket Shop' is having a sale.

How much did this jacket cost before the sale?

The jacket costs 25% less in the sale, so it is

> 100% − 25% = 75% of the original price
>
> 75% of the original price = £60
>
> 1% of the original price = $\frac{£60}{75}$ = £0.80
>
> 100% of the original price = £0.80 × 100 = £80

**The jacket cost £80 before the sale.**

*Another jacket costs £81 in the sale. What did it cost before the sale?*

## 1: Fractions, decimals and percentages

**1** The price of a monthly train ticket increases by 12% to £84.

What did it cost before the increase?

**2** Richard is given a 16% pay rise. His salary is £29 232.

What was his salary before the rise?

**3** Sarah sees this sign.

**COMPUTER STOREHOUSE**
**SUMMER SALE**
**30% off everything**

She pays £596.40 for a computer in the sale.

a) What was the original price of the computer?

b) How much does she save?

**4** Marco is buying some furniture for his new flat.

£94

£235

£119.85

The prices include 17.5% VAT.

Work out the price of each item excluding VAT.

**5** Ryan buys a washing machine. He pays 12 monthly instalments of £34.

a) How much does he pay in total?

b) Ryan had to pay 20% more than the original shop price because he paid by instalments.

Work out the original shop price.

**6** Zoe sells her car for £5170. She makes a 6% loss on the original price.

How much did Zoe originally pay for the car?

> Find a list of second hand car prices.
>
> Work out the percentage yearly loss in value for different cars.

## 1: Fractions, decimals and percentages

# Percentage problems

Josh works for a charity that wants to buy this computer.

The charity does not pay VAT.

He works out how much the charity pays like this:

117.5% is 940

1% is $\dfrac{940}{117.5}$

100% is $\dfrac{940}{117.5} \times 100 = 800$

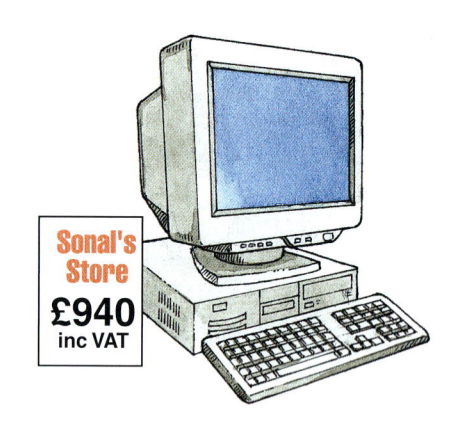

**?** *How much does the charity pay for the computer?*

Josh goes to another store and buys this printer. There is a special promotion today.

He pays £170 for the printer.

> The usual price is 100% and the discount is 15% so he pays 100% − 15% = 85%

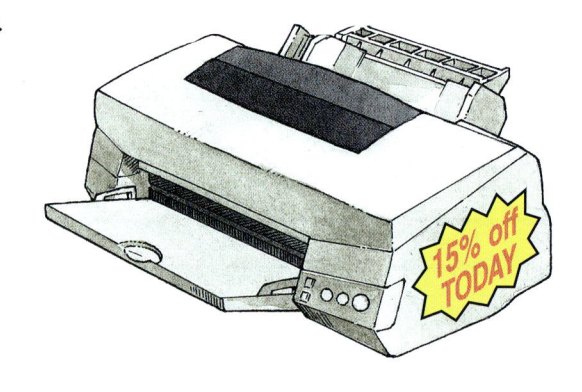

Josh works out the usual price like this:

85% is 170

1% is $\dfrac{170}{85}$

100% is $\dfrac{170}{85} \times 100 = 200$

**?** *How much does Josh save by buying it on the special promotion?*

## 1: Fractions, decimals and percentages

**1** These prices are inclusive of VAT at 17.5%.

Work out the price exclusive of VAT.

a) £188   b) £150   c) £69.99

**2** Jenna has just got a 4% pay rise. Her salary is now £13 000.

a) What was her salary before the rise?

b) Twelve months later she gets another 4% rise.

What is her salary after this rise?

**3**

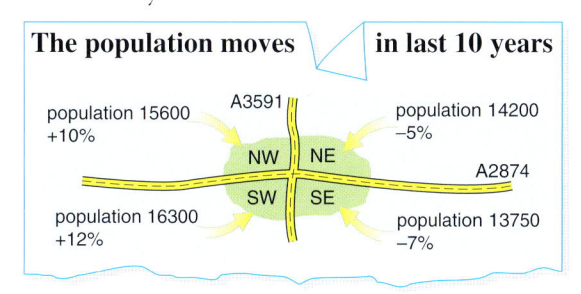

**The population moves           in last 10 years**

- NW: population 15600 +10%
- NE: population 14200 −5%
- SW: population 16300 +12%
- SE: population 13750 −7%

A3591, A2874

The newspaper gives the present population of each quarter of the town. The percentage change that has taken place in the last 10 years is also given.

a) What is the missing word in the headline? Choose North, South, East or West.

b) Work out the population of each quarter as it was 10 years ago.

c) Work out the percentage change in the number of people living North of the A2874.

**4** Henry is a salesman. His sales are £350 000 this year.

This is 25% more than 3 years ago and 85% more than 7 years ago.

Work out his sales figures (to the nearest thousand pounds)

a) three years ago

b) seven years ago.

c) His target sales for next year is an 8% increase, with a further 8% increase the year after.

Work out his target sales figures for the next two years.

## 1: Fractions, decimals and percentages

# Finishing off

**Now that you have finished this chapter you should be able to**

★ multiply and divide fractions

★ work out percentage problems when you don't know the original amount

★ do rough calculations to check your answer

★ calculate the original price after a percentage increase or decrease.

Use the questions in the next exercise to check that you understand everything.

## Mixed exercise

**1** Work out

a) $2\frac{7}{8} + 1\frac{3}{4}$   b) $5\frac{1}{16} - 4\frac{1}{4}$   c) $5\frac{2}{3} - 1\frac{1}{6}$   d) $2\frac{4}{5} + 6\frac{7}{10}$

e) $\frac{1}{4} \times 2\frac{2}{3}$   f) $6\frac{3}{4} \times 1\frac{1}{2}$   g) $4\frac{5}{8} \times 1\frac{1}{4}$   h) $1\frac{1}{3} \times 3\frac{3}{4}$

i) $1\frac{1}{2} \div 2$   j) $2\frac{3}{4} \div \frac{1}{4}$   k) $1\frac{1}{8} \div 4\frac{1}{2}$   l) $6\frac{1}{4} \div 1\frac{2}{3}$

**2** Work out

a) $5\frac{1}{2} - 3\frac{3}{4} - 1\frac{1}{8}$   b) $(\frac{2}{3} - \frac{1}{6}) \div 5$   c) $6 \div (\frac{1}{2} + \frac{1}{4})$

d) $2\frac{1}{2} \times 1\frac{3}{5} \times 1\frac{1}{4}$   e) $2\frac{1}{4} + 3\frac{2}{3} + 4\frac{1}{2}$   f) $(4\frac{1}{2} - 1\frac{1}{6}) \div \frac{3}{4}$

**3** Work out

a) $4.2 + 3.97$   b) $0.3 \times 0.3$   c) $2 \div 0.4$   d) $1.65 \times 2.4$

e) $53.2 \div 10$   f) $39 \times 100$   g) $0.74 \times 1000$   h) $13.7 \div 1000$

**4**

Geoff's Gadgets — £199

Steve's Store 20% off — £245

Tessa's TV's $\frac{1}{6}$ off — £235

a) Which television is the cheapest and how much does it cost?

b) The television in Geoff's Gadgets is inclusive of VAT at 17.5%. Work out the price exclusive of VAT.

## 1: Fractions, decimals and percentages

**Mixed exercise**

**5** At a rugby match two thirds of the crowd are home supporters, one quarter are away supporters and the remainder are neutrals.

There are 1250 neutrals. How large is the crowd?

**6** Ranjit is looking at how prices have changed in the last 3 years.

| Item | Price now | % change over last 3 years |
|---|---|---|
| House | £75 200 | up 14% |
| Car | £11 400 | up 11% |
| Computer | £1199 | down 30% |
| Calculator | £7.50 | down 25% |

Work out the prices 3 years ago giving your answers to 3 significant figures.

**7** Isabel is testing children's mental arithmetic skills using a 20 question test.

a) The pass mark is 16. What percentage is this?

Isabel tests children in 3 schools and gets these results.

| School | Number passing | Number tested |
|---|---|---|
| Greenway | 185 | 279 |
| Eastwood | 149 | 237 |
| Parkside | 261 | 364 |

b) Which school has the highest pass rate?

c) Which school has the lowest pass rate?

**8** Lynn wants to take a taxi from the centre of London to Heathrow Airport (a distance of 30 miles). She finds out the fares from two taxi companies.

Calculate the cost of the journey using each of the companies.

**TINA'S TAXIS** £4 plus 30p per mile

**Colin's Cabs** 45p per mile

**9** All of these prices are excluding VAT. Work out the cost including VAT at 17.5%.

a) £480  b) £350  c) £6900  d) £7224  e) £12 226

### 1: Fractions, decimals and percentages

**Mixed exercise**

**10** Calculate the percentage profit or loss on an item

a) bought for £30 and sold for £42

b) bought for £1230 and sold for £1414.50

c) bought for £48 and sold for £36

d) bought for £75 and sold for £52.50.

**11** Work out the selling price of an item

a) bought for £620 and sold at a 15% profit

b) bought for £35 and sold at a 22% profit

c) bought for £224 and sold at a 16% loss

d) bought for £8246 and sold at a 11% loss.

**12** Peter is a market trader. He buys 20 shirts at £12 each. He prices them so that he makes a 40% profit on each shirt.

a) What price does he charge for each shirt?

b) He sells 14 of them at this price and the remainder are sold for £10.

How much does he take altogether?

c) What is his total profit?

d) Work out this profit as a percentage of his total outlay.

**13** Stephen gets a 5% pay rise. His new salary is £17 282.

What was his salary before?

**14** Work out the original price of each of these items.

SALE ! Everything has been reduced by 25%!

SALE PRICE £261.75

SALE PRICE £63

SALE PRICE £149.25

**15** a) Emma sells her house for £146 400 at a 22% profit. How much did she pay for it?

b) Peter sells a jacket for £120 at a 40% profit.

How much did he pay for it?

c) Julie sells her exercise bike for £64 at a 20% loss.

How much did she pay for it?

## 1: Fractions, decimals and percentages

**Mixed exercise**

**16** The local travel agent has increased the price of all holidays by 6% since last year.

a) Ravi paid £900 for his holiday last year.

How much is the same holiday this year?

b) June paid £420 for her holiday last year.

How much is the same holiday this year?

c) Hannah goes on the same holiday she went on last year.

The new cost of her holiday is £689.

What did she pay last year?

d) David goes on the same holiday he went on last year.

It costs him £60 more than it did last year.

(i) What did he pay last year?

(ii) What does he pay this year?

**17** Kerry is a telephone salesperson. She earns a basic salary of £200 per month and gets 12% commission on her sales.

These are her sales figures for January to June.

| January | February | March | April | May | June |
|---------|----------|-------|-------|-----|------|
| £4000   | £4250    | £3600 | £4800 | £4400 | £5200 |

a) Calculate her total pay for each month.

b) How much does she earn in total over the six months?

c) Kerry's company offers her the chance to change the way she is paid. She can choose from:

    **Option A:** A basic salary of £750 per month (no commission)

    **Option B:** A basic salary of £150 per month plus 15% commission.

Use the sales figures above to work out how much she would have been paid using

(i) Option A

(ii) Option B

d) What would you advise Kerry to choose?

# Two

# Formulae and equations

**Before you start this chapter you should be able to**

★ carry out calculations in the correct order

★ multiply and divide using negative numbers

★ add and subtract using negative numbers

★ understand the meanings of $2x$ and $x^2$

★ expand brackets

★ collect together like terms

★ solve simple equations

★ read information from a graph.

Use the following questions to check that you still remember these topics.

## Reminder

- The correct order for a calculation is

    Brackets, Indices, Divide, Multiply, Add, Subtract (BIDMAS).

- When adding and subtracting with negative numbers a number line is helpful.

- When multiplying or dividing with negative numbers, or when there are two signs before a number,

    + with + gives +      + with − gives −      − with − gives +

- $2x$ means $2 \times x$ or $x + x$, and $x^2$ means $x \times x$.

### Revision exercise

*Always start by copying the expression in the question, then work down your page so you can see where each term comes from.*

**1** Simplify each of these, remembering to work in the correct order.

a) $6 + 3 \times 2$

b) $14 + 3 \times 3 - 2^3$

c) $2 \times 3^2$

d) $2(4x - 7)$ when $x = 3$

e) $2(x - 4)$ when $x = 4$

f) $5n^2$ when $n = 2$

g) $x \div 3$ when $x = 0$

h) $(2n)^2$ when $n = 5$

i) $(10 + 2)(5 - 5)$

j) $(x + 4)(x + 1)$ when $x - 3$

## 2: Formulae and equations

**Revision exercise**

**2** Simplify these.
  a) $2 - 12$
  b) $5 + 13 - 20$
  c) $-2 - 6 - 9$
  d) $8x - 12x$
  e) $4m + 2m - 5m$
  f) $3c - 4c + c$

**3** Write these as briefly as possible.
  a) $3 \times a + 6 \times b$
  b) $12 - 6 \times x$
  c) $3 \times 5 \times c$
  d) $3 \times n \times n \times n \times n$

**4** Simplify these.
  a) $5 \times (-3)$
  b) $-2 \times (-2)$
  c) $12 \div (-4)$
  d) $-12 \div 4$
  e) $6 + (-2)$
  f) $8 - (-5)$
  g) $3y \times (-2)$
  h) $-4m \div (-2)$

**5** Solve each of these equations to find the unknown.
  a) $17 + x = 98$
  b) $12x = 84$
  c) $y - 46 = 23$
  d) $5x + 12 = 99$
  e) $34.8 - x = 25.8$
  f) $2x + 7 = 0$

**6** Expand these brackets.
  a) $2(n - 3)$
  b) $(a - b + c) \times 7$
  c) $(5a + 6) \times 4$
  d) $3(4y - 2z)$

**7** Collect together the like terms and then simplify these expressions.
  a) $2n + 3 + 3n - 4$
  b) $6y - 5 + 10 - 6y$

**8** Expand the brackets and then simplify these by collecting like terms.
  a) $2(x + 8) + 1$
  b) $12 + (5 + y)$
  c) $3 + 2(2n - 1)$
  d) $7 - 3(a - 2)$

**9** These graphs have been drawn by using equations to work out the co-ordinates. One is $y = x$, one is $y = x^2$ and the other is $y = \dfrac{1}{x}$.

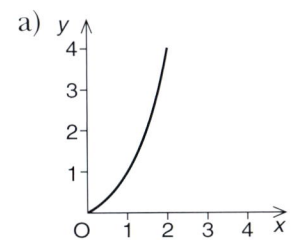

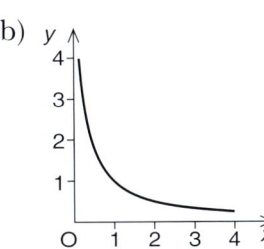

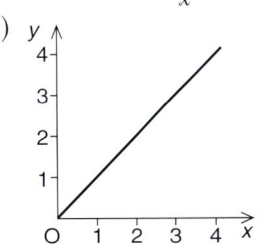

  a), b), c) [graphs]

Which equation fits which graph? Give reasons for your answers.

## 2: Formulae and equations

# Making up formulae

Leanne is counting off the days before her summer holiday.

**?** *How many days does she have to wait if there are 14 weeks to go before then?*

Suppose $W$ stands for the number of weeks and $D$ stands for the number of days.

You can write a formula to work out the value of $D$ for different values of $W$:

$$D = 7 \times W \quad \text{or} \quad D = 7W$$

*$D$ is called the **subject** of this formula*

When Leanne has 10 weeks to wait, $W = 10$; the number of days Leanne has still to cross off is:

$$D = 7 \times 10$$
$$= 70$$

You can show $D$ and $W$ on a graph.

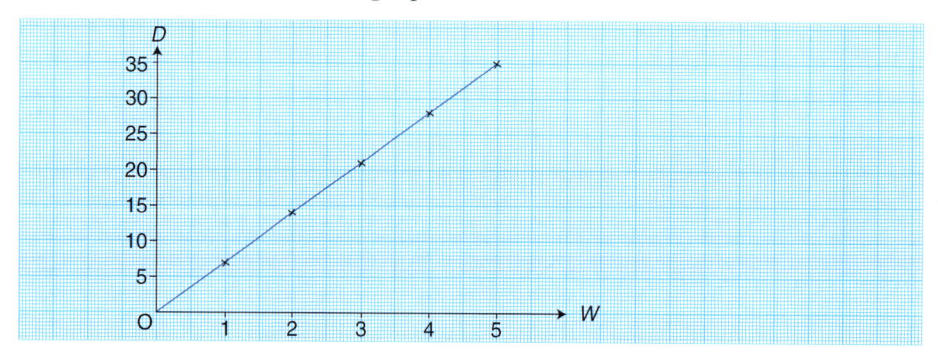

In a formula or an equation, the letter always stands for a number.

**?** *Leanne says to herself 'A week is 7 days so $W = 7D$.'*

*Use $D = 14$ to show this is wrong.*

To make $W$ the subject of $D = 7W$, write $7W = D$.

Divide both sides by 7, $W = \dfrac{D}{7}$

## 2: Formulae and equations

**1** Justin can't quite remember a formula. He knows it is one out of:

1. $H = 40 + 10t$  2. $H = 40 - 10t$
3. $H = 10 + 40t$  4. $H = 10 - 40t$

a) Use each of these formulae to find $H$ when $t = 2$.

b) Justin knows that $H = 20$ when $t = 2$. Which formula is correct?

**2** Today is Alex's birthday. Her age is $y$ years.

a) Write down a formula for $m$, Alex's age in months.

b) Check your formula by letting $y = 10$.

c) Write down a formula to give $y$ in terms of $m$.

**3** A table is 2 m long. How would you find its length in centimetres?

a) Using $C$ as the number of centimetres and $M$ as the number of metres, write a formula of the form $C = \ldots \times M$.

b) Now write the formula with $M$ as the subject, $M = \ldots$

**4** The graph shows how many miles, $m$, a car can travel on $l$ litres of petrol.

a) How many miles does the car do on 1 litre?

b) Write a formula giving $m$ in terms of $l$.

c) Rewrite your formula with $l$ as subject.

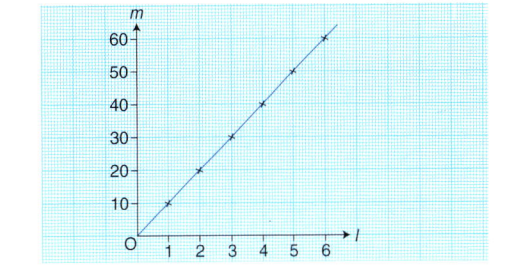

**5** Write down formulae connecting

a) (i) the price in pence, $P$, of $A$ apples costing 30p each

   (ii) the number of apples, $A$, you can buy for $P$ pence.

b) (i) the number of centimetres, $C$, in $I$ inches

   (ii) the number of inches, $I$, in $C$ centimetres.

**6** The temperature at dawn is $D\,°C$.

The midday temperature, $M$, is 12°C greater.

a) Write a formula giving $M$ in terms of $D$.

b) Write a formula giving $D$ in terms of $M$.

---

Find three formulae which are commonly used in other subjects. They might be in words. If so, rewrite them using letters.

State clearly what the letters represent in all your formulae.

In each case write the formula in more than one way, using a different subject.

## 2: Formulae and equations

# Working with unknowns

*Try this out on a few people.*

> 'Think of a number,......add 1,......multiply by 2,......add 4,
> ......divide by 2,......subtract the number you first thought of,
> ......what is your answer?'

*What happens? Why?*

Justin uses algebra to work out what is happening.

Since he doesn't know what number a person will think of, he calls it $n$.

He writes this:

| Think of a number | $n$ |
| --- | --- |
| Add 1 | $n+1$ |
| Multiply by 2 | $2(n+1) = 2n+2$ |
| Add 4 | $2n+6$ |
| Divide by 2 | $n+3$ |
| Subtract the number ($n$) | $3$ |

*Justin divides both these terms by 2*

The answer is 3. It does not depend on the value of $n$. So whatever number the person thinks of, the answer will be 3.

This sets Justin thinking. He invents this new trick.

> 'Think of a number,......add 2,......multiply by 4,......add 2,
> ......divide by 2,......subtract 5,......tell me the answer.'

When people tell him the answer, Justin tells them the number they first thought of. He gets it right every time.

*How does Justin know what number the person first thought of?*

*Find the rule that links the answer to the number you first thought of.*

*How can you use algebra to prove that your idea is right?*

## 2: Formulae and equations

**1** This question revises the use of brackets. Check that you understand it before you go on to the other questions.

Write these without brackets and simplify them where possible.

a) $4(n + 2)$     b) $5(m - 2)$     c) $7(4 + 2x)$

d) $8(2 - 3y)$     e) $3(2 - x) + 5$     f) $10(n + 4) - 30$

g) $2(3x + 6) - 12$     h) $15 + 3(x - 5)$     i) $5(5x + 2) - 16$

j) $10 - 2(n + 1)$     k) $22 - 3(a + 5)$     l) $4 - 2(x - 1)$

**2** Work through these using $n$ for the number. For each one say what answer you would expect.

a) 'Think of a number,......subtract 1,......add 4,......subtract the number you first thought of.'

b) 'Think of a number,......add 2,......multiply by 2,......subtract 2, ......divide by 2,......subtract the number you first thought of.'

c) 'Think of a number,......multiply by 10,......add 2,...... multiply by 3, ......subtract 6,......divide by the number you first thought of.'

d) 'Think of a number,......subtract 1,......multiply by 5,......add 5, ......divide by the number you first thought of.'

Check your answers by trying the tricks out on someone.

**3** Try this with a few numbers:

'Think of a number,......add 10,......multiply by 10,......subtract 100.'

a) Describe how you can 'mind read' the number first thought of once you are told the answer.

b) Use algebra to show how this works.

**4** Work through these as in question 3, then explain how to work out the original number. Try them out to check your answers.

a) 'Think of a number,......subtract 1,......multiply by 4,......add 8, ......divide by 4.'

b) 'Think of a number,......add 3,...... multiply by 3,...... subtract 3, ......divide by 3.'

c) 'Think of a number,......subtract 1,......multiply by 10,...... subtract 10, ...... divide by 10.'

Make up some 'Think of a number......' puzzles.
Check them through, using algebra, and make sure they don't involve any very difficult mental arithmetic.

### 2: Formulae and equations

# Using equations to solve problems

**?** Which of these telephone companies is cheaper if you make just a few calls a month?

Which is cheaper if you make a lot of calls each month?

How do you decide which is cheaper for you?

**VENUS TELECOM**
No Rental
No Standing charge
JUST 5p per minute for calls!

**Connect!**
Monthly rental £12
Calls ONLY 2p per minute!

You can do this using trial and error, or you can gain a clearer picture of the situation by using algebra.

The charges depend on the number of minutes that your calls last each month. Call the number of minutes $m$.

*It is important to be clear about what the letters and expressions stand for, and what units they are in.*

The amount in pence charged by Venus is then  $5m$.

The amount in pence charged by Connect is  $1200 + 2m$.

**?** What does the 1200 represent in the expression for Connect's charges?

To find the value of $m$ for which both companies charge the same amount, you form an equation and then solve it.

| The companies charge the same amount when | $5m = 1200 + 2m$ |
| Subtract $2m$ from both sides | $3m = 1200$ |
| Divide both sides by 3 | $m = 400$ |

*This is an equation in $m$.*

**The companies charge the same amount when  $m = 400$.**

**?** Nick's calls last about 600 minutes each month. Which company should he use?

Catherine's calls are usually under 300 minutes each month. Which company should she use?

There are many situations in which forming and solving an equation is helpful.

**?** A chocolate bar is marked '25% extra free'. Its weight is 250 g.

How would you form an equation to find the weight of a normal bar?

## 2: Formulae and equations

**1** This question revises how to solve equations. Check that you understand it before you move on to the next questions.

Solve these equations.

a) $2x + 5 = 25$    b) $3x - 8 = 10$    c) $4x - 9 = 2x + 11$

d) $x + 8 = -x + 14$    e) $3x + 4 + x = 3x + 7$    f) $5(x + 2) = 2x + 22$

g) $5(x + 4) = 4(x + 5) + 3$    h) $7(x + 7) = 49$    i) $3(x - 2) = 4x - 7$

**2** For each of the following situations
(i) form an equation in the unknown quantity given
(ii) solve the equation
(iii) check your answer.

a) Fanzia goes shopping with £100 in her purse. She buys $x$ CDs at £12 each and still has £16 left.

b) In one season, Totnes Wanderers Football Club scores 72 points. They win $w$ matches (3 points each), draw 6 (1 point each) and lose the rest (0 points).

c) The length of a field is 3 times its width of $w$ metres. The perimeter is 600 m.

d) The largest angle of a triangle is 4 times the size of the smallest angle, $A°$. The third angle is 60°.

e) Halley's present age is $y$ years. In 24 years time he will be 3 times as old as he is now.

**3** Sara is finding out about monthly charges for using the Internet. She has written these notes.

> A-SERVE
> charges a flat rate of £5 per month
> BEELINE
> charges £5 for first 5 hours and £2 for each hour over 5
> COMIC
> charges £9 for first 5 hours and then £1·50 for each hour over 5

Sara intends to surf the Internet for more than 5 hours. Suppose $x$ stands for the number of extra hours (above 5) that she spends.

a) Write down expressions for the amount Sara could expect to be charged by each company.

Which is cheapest when $x = 1$?
Which is cheapest when $x = 10$?

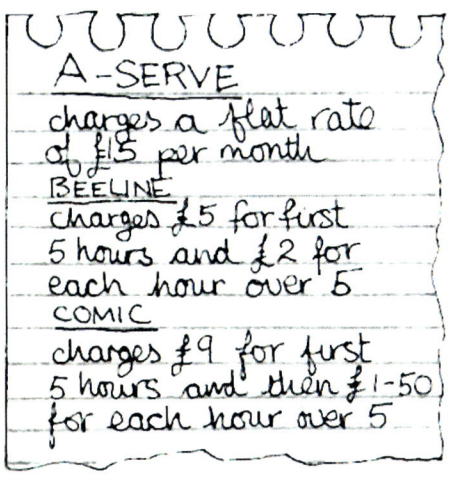

b) For a certain value of $x$, A-serve and Beeline must charge the same. Put those two expressions equal to each other, and solve the resulting equation to find this value of $x$.

c) Repeat b) to find a value of $x$ for which Beeline and Comic charge the same.

d) Do the same for A-serve and Comic.

e) What advice would you give Sara?

## 2: Formulae and equations

# Using graphs to solve equations

So far you have used algebra to solve the equations you have met, but this is not always possible. Some equations are just too complicated. In that case you can always use a graph. The method is shown below for the equation

$$x^3 - 4x + 1 = 0$$

Start by drawing the graph of $y = x^3 - 4x + 1$.

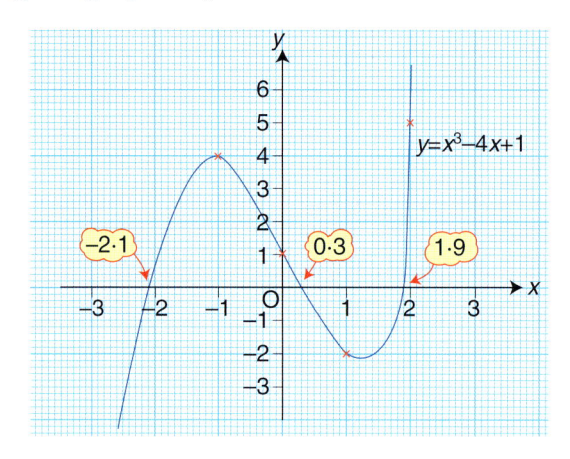

*Be sure to show the places where the graph crosses the x axis*

You can see that this graph crosses the $x$ axis at three places. Written to 1 decimal place, these are –2.1, 0.3 and 1.9.

**?** *Why do these give the solution of the equation?*

**The solution of the equation is $x = -2.1$, 0.3 or 1.9 (approximately).**

**?** *What are the advantages and disadvantages of this method of solving equations?*

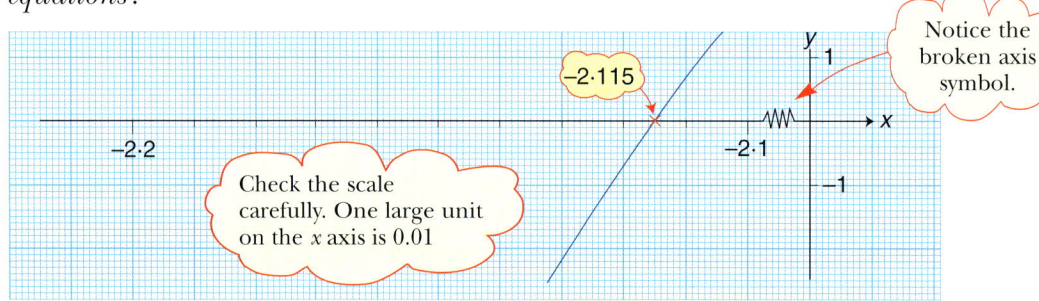

*Notice the broken axis symbol.*

*Check the scale carefully. One large unit on the x axis is 0.01*

You can find the solution more accurately by drawing the graph to a larger scale, or by drawing close-ups of the places where it crosses the $x$ axis.

This diagram is a close-up of part of $y = x^3 - 4x + 1$. It shows that –2.115 is more accurate than –2.1. The main problem is that doing this takes a long time.

**?** *How would you use the upper graph on this page to solve the equation*
$$x^3 - 4x + 1 = 3?$$

28

## 2: Formulae and equations

**1** The graphs in this question are all drawn to different scales. State the value of $x$ where each curve crosses the $x$ axis.

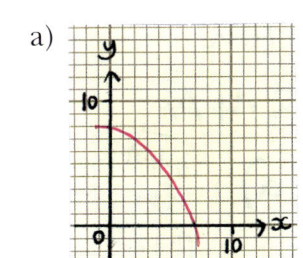

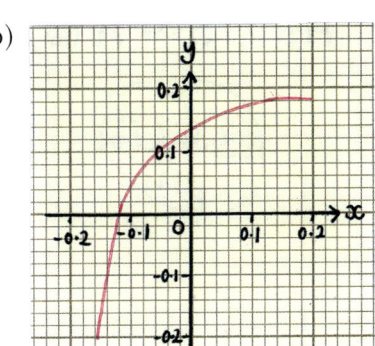

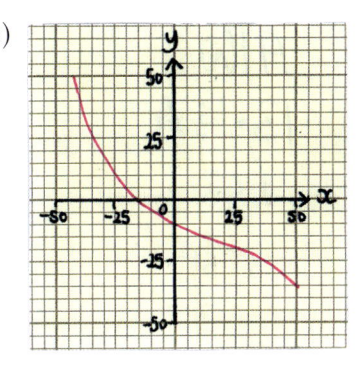

a)   b)   c)

**2** a) Draw the graph of $y = x^3 - 5x - 2$ taking values of $x$ from $-3$ to $+3$.

b) Use your graph to solve the equations
   (i) $x^3 - 5x - 2 = 0$
   (ii) $x^3 - 5x - 2 = 4$

**3** a) Draw the graph of $y = \dfrac{6}{x}$ taking values of $x$ from 1 to 12.

b) Use your graph to solve the equation $\dfrac{6}{x} = 4$.

c) Use algebra to solve the equation $\dfrac{6}{x} = 4$.

d) Which do you think is the better method, drawing a graph or using algebra?

**4** a) Draw the graph of $y = x^3 - 2x - 1$ for values of $x$ from 0 to 2.

Estimate the solution as accurately as you can. You should find that it is between 1.5 and 1.7.

b) Draw a close-up of the part of the graph between $x = 1.5$ and $x = 1.7$ to get a more accurate estimate of the solution.

c) Draw a further close-up to get an even more accurate estimate.
To how many decimal places do you think your answer is accurate?

---

You can use a graphic calculator and its zoom function to find the solution of an equation.

Find out how to do this, and use it to solve the equation $x^3 - 5x - 2 = 0$ to 2 decimal places. Write simple instructions telling a friend how to do this.

## 2: Formulae and equations

# Trial and improvement

You have seen how to use graphs to solve equations. If you want to be more accurate you can draw close-ups of the places where they cross the x axis but this takes a long time. However if you keep the idea of the graph in your mind, you don't actually need to draw the close-ups.

Suzanne needs to solve the equation $x^3 - 3x^2 - 1 = 0$.
Here is some of her working.

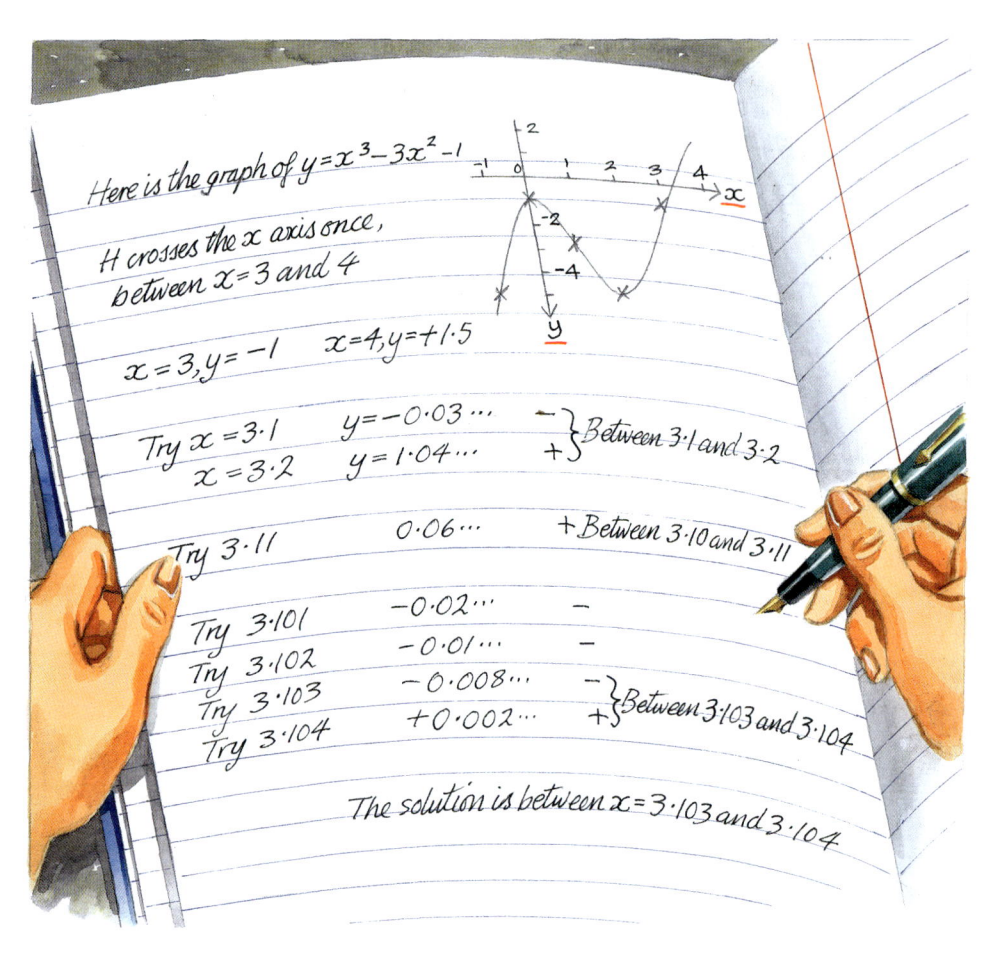

*Here is the graph of $y = x^3 - 3x^2 - 1$*
*It crosses the x axis once, between $x = 3$ and $4$*

$x = 3, y = -1$  $x = 4, y = +15$

Try $x = 3.1$  $y = -0.03...$  — } Between 3.1 and 3.2
    $x = 3.2$  $y = 1.04...$  +

Try 3.11   $0.06...$   + Between 3.10 and 3.11

Try 3.101  $-0.02...$  —
Try 3.102  $-0.01...$  —
Try 3.103  $-0.008...$ — } Between 3.103 and 3.104
Try 3.104  $+0.002...$ +

*The solution is between $x = 3.103$ and $3.104$*

**?** Why does Suzanne not write these numbers to more decimal places? What is the answer correct to 2 decimal places?

This method is called **trial and improvement**.

**?** You can find $\sqrt[3]{10}$ by solving the equation $x^3 = 10$. Where do you start?
How many calculations do you then need to do to find the answer correct to 3 decimal places?

**?** You could also solve Suzanne's equation by writing it as $x^3 - 3x^2 = 1$.
What is the value of $x^3 - 3x^2$ when    a) $x = 3$?    b) $x = 4$?
What would you do next?

## 2: Formulae and equations

**1** For each of these equations, find two consecutive whole numbers between which the solution must lie. Then find the solutions of the equations correct to 1 decimal place.

a) $x^3 = 4$
b) $x^3 - x - 80 = 0$
c) $4x^3 - 157 = 0$
d) $2x^3 + 3x - 1500 = 0$
e) $x + \dfrac{1}{x} = 5$
f) $x^3 - 7x^2 + 8x - 3 = 0$

**2** Use trial and improvement to solve $x^3 - 7x - 9 = 0$ to 3 decimal places.

**3** Use trial and improvement to solve $x^2 - \dfrac{1}{x} = 1$ to 3 decimal places.

**4** Matt is designing a carton for a special party fruit drink. The manufacturer wants the carton to hold 4 litres and Matt decides to make it the shape in the diagram.

a) Show that the volume is $(x^3 + x^2)$ cm$^3$.

b) Explain why $x^3 + x^2 = 4000$.

c) Solve the equation $x^3 + x^2 - 4000 = 0$, and so find the lengths of the sides to the nearest millimetre.

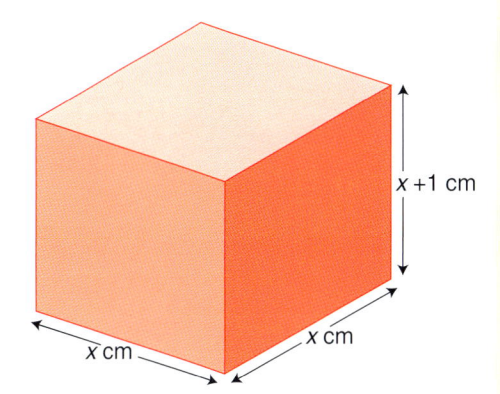

*x +1 cm*
*x cm*
*x cm*

**5** Majid throws a tennis ball over his parents' garage. The graph shows its trajectory. As you can see it just clears the roof on the far side. The equation of the path of the ball is $y = 1 + 2x - 0.2x^2$ and the height of the garage is 3 m.

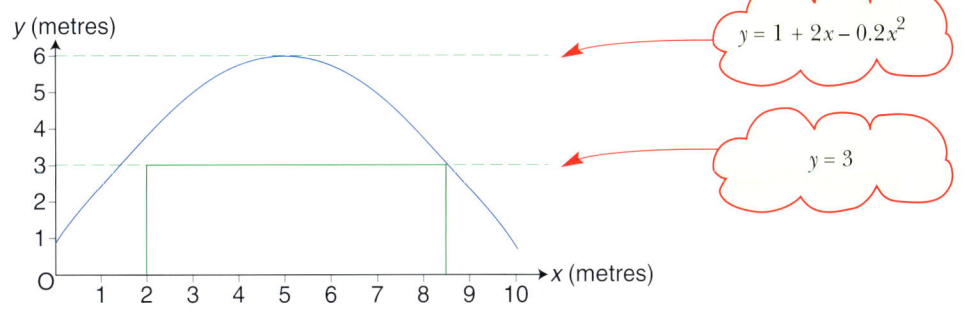

*y = 1 + 2x − 0.2x²*
*y = 3*

Use a trial and improvement method to find the length of the garage to the nearest centimetre.

Find out how using a spreadsheet can help you to solve an equation by trial and improvement. Choose an equation and solve it by this method. The spreadsheet printouts should form part of your answer.

## 2: Formulae and equations

# Rearranging a formula

In this section you revise the work on rearranging formulae that you covered in Book 1.

## Reminder

- When you rearrange a formula you must do the same to each side of it, just as you do when you are solving an equation.
- Think about the order of operations on $x$ in the formula. You must 'undo' them in reverse order.
- Isolate squared and square root terms before dealing with them.

### Example

Make $x$ the subject of these formulae.

a) $\dfrac{ax + b}{c} = d$

b) $\sqrt{\dfrac{x}{w}} + y = z$

### Solution

a) $\dfrac{ax + b}{c} = d$

$ax + b = cd$  *(Multiply both sides by $c$.)*

$ax = cd - b$  *(Subtract $b$ from both sides.)*

$x = \dfrac{cd - b}{a}$  *(Divide both sides by $a$.)*

b) $\sqrt{\dfrac{x}{w}} + y = z$

$\sqrt{\dfrac{x}{w}} = z - y$  *(Isolate the square root term before squaring.)*

$\dfrac{x}{w} = (z - y)^2$  *(Notice the use of brackets.)*

$x = w(z - y)^2$

## Formulae in which the new subject appears more than once

Suppose you want to make $x$ the subject of the formula
$$ax - b = cx + d.$$

Notice that there are two $x$ terms in the formula. To make $x$ the subject, you need first to collect the $x$ terms together. It is a bit like solving an equation such as
$$5x - 1 = 3x + 7.$$

**Solving the equation**

$5x - 1 = 3x + 7$

$5x = 3x + 7 + 1$  *(Add 1 to each side.)*

$5x - 3x = 8$  *(Subtract $3x$ from each side.)*

$2x = 8$

$x = 4$  *(Divide both sides by 2.)*

**Rearranging the formula**

$ax - b = cx + d$

$ax = cx + b + d$  *(Add $b$ to each side.)*

$ax - cx = b + d$  *(Subtract $cx$ from each side.)*

$(a - c)x = b + d$  *(Notice the use of brackets to create a single $x$ term.)*

$x = \dfrac{b + d}{a - c}$  *(Divide both sides by $a - c$.)*

## 2: Formulae and equations

**1** Make $x$ the subject of each of these formulae.

a) $wx + y = z$  
b) $s - x = t$  
c) $\dfrac{x}{a} - b = c$  
d) $\dfrac{y}{x} = z$  
e) $\dfrac{p - x}{q} = r$  
f) $\dfrac{r}{s - x} = t$  
g) $\dfrac{x^2 + a}{b} = c$  
h) $\sqrt{cx - d} = e$  
i) $p - \sqrt{\dfrac{x}{q}} = r$  
j) $\dfrac{f}{(x - h)^2} = g$  
k) $\dfrac{w}{\sqrt{x + y}} = z$  
l) $a - \dfrac{b}{x^2} = c$

**2** Make $x$ the subject of each of these formulae.

a) $ax + b = cx + d$  
b) $px - q = qx$  
c) $xy + z = y - xz$  
d) $\dfrac{cx + d}{e} = x$  
e) $\dfrac{x - a}{b} = \dfrac{x + b}{c}$  
f) $\dfrac{x + p}{x - q} = p$  
g) $\dfrac{x}{s} + \dfrac{x}{t} = r$  
h) $\dfrac{a}{x} + \dfrac{b}{ax} = b$  
i) $\sqrt{\dfrac{x - y}{x + w}} = z$  
j) $\dfrac{\sqrt{x^2 + p}}{x} = p$

**3** The formula for the surface area of a cone is
$A = \pi r^2 + \pi r \sqrt{h^2 + r^2}$.

Make $h$ the subject of the formula.

**4** A flower bed is cut out of a square piece of lawn.

The unshaded part of the diagram shows the part of the lawn which is cut away. The boundary of the flower bed consists of four circular quadrants, each of radius $r$.

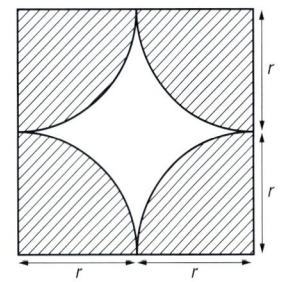

a) Show that the unshaded area, $A$, is given by the formula
$A = 4r^2 - \pi r^2$.

b) Rearrange this formula to obtain $r$ in terms of $A$.

c) Hence, or otherwise, find the value of $r$ required to make a flower bed of area 25 m².

*NEAB*

## 2: Formulae and equations

# Finishing off

**Now that you have finished this chapter you should**

★ be able to substitute numbers into formulae

★ be confident about working with brackets

★ be able to use equations to solve problems

★ solve equations graphically

★ use trial and improvement methods to solve equations.

Use the questions in the next exercise to check that you understand everything.

**Mixed exercise**

**1** Write each of these expressions without brackets and then simplify it.

a) $3(x + 2)$
b) $5(x + 2) + 2(x + 3)$
c) $6 + 2(x - 3)$
d) $12(m + 5) - 34$
e) $21 + 3(x - 7)$
f) $13 - 4(n + 1)$
g) $21 - 5(x - 1)$
h) $6 - 3(2y - 1)$
i) $11 - 6(3x - 5)$

**2** Tariq throws a snowball. At any point during its flight the snowball's height, $y$ metres, is related to its horizontal distance, $x$ metres, by the equation

$$y = x - \frac{1}{6}x^2 + \frac{1}{2}.$$

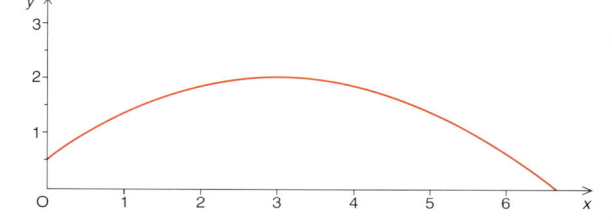

The graph of this equation is shown in the diagram.

a) Use the equation to find $y$ when $x = 0$, 3 and 7.

b) What do your answers to a) tell you about the height of the snowball at each of these points?

c) Tariq is aiming to hit a fence 7 m away. Does he succeed?

**3** Lloyd thinks of a number $x$.

He multiplies it by 9 and subtracts 12 and makes the answer 33.

Make an equation and solve it to find Lloyd's number.

## 2: Formulae and equations

**Mixed exercise**

**4** Ching asks Jo to think of a number, then to subtract 1, multiply by 4, and finally subtract 10.

   a) Calling Jo's number $n$, find an expression for her answer in terms of $n$.

   b) Jo says her answer is twice the number she first thought of.

   Make an equation for $n$ and solve it to find Jo's number.

**5** Morag buys a coat that has been reduced by 20%. The original price was £$C$.

   a) Write 20% as a decimal, and so write an expression for the amount by which the coat was reduced, in terms of $C$.

   b) Find the sale price of the coat in terms of $C$.

   c) Morag actually paid £48 in the sale. Make an equation and solve it to find $C$.

**6** When Grandma Jones visits her family, her return train fare is £15.

   a) Write down an expression for the cost of $x$ return journeys.

   b) Grandma Jones could buy a railcard for £20. With the railcard the fare is reduced to $\frac{2}{3}$ of the usual price for a period of one year.

   Write an expression for the total cost of $x$ journeys using the railcard.

   c) Use your answers to a) and b) to find a value of $x$ which will make the costs equal.

   d) How many times a year does Grandma Jones need to visit her family to make it worth buying a railcard?

**7** a) Draw the graph of
   $y = x^3 - 6x^2 + 11x - 6$
   for values of $x$ between 0 and 4.

   b) Solve the equation
   $x^3 - 6x^2 + 11x - 6 = 0$

   c) Solve the equation
   $x^3 - 6x^2 + 11x - 6 = 1$.

**8** a) Draw the graph of $y = \frac{12}{x} - x^2$ for values of $x$ between 1 and 4.

   b) Use your graph to solve the equation $\frac{12}{x} - x^2 = 0$.

   Give your answer to 1 decimal place.

   c) Use your calculator to find $\sqrt[3]{12}$.

   d) Explain why your answers to parts b) and c) are the same.

**9** Use trial and improvement methods to find the solution of the equation.

   a) $x^3 - 2x^2 - 1 = 0$

   b) $x - \frac{1}{x^2} - 2 = 0$

# Three

# Triangles and polygons

**Before you start this chapter you should**

★ be able to classify triangles

★ recognise different types of quadrilateral

★ be able to find pairs of equal angles where two lines cross or where a line intersects parallel lines

★ know that angles round a point add up to 360° and angles on a straight line add up to 180°

★ know that the angle sum of a triangle is 180°.

Use the following questions to check that you still remember these topics.

## Reminder

- Where two lines intersect, **opposite angles** are equal.

- Where a line intersects with two parallel lines, **corresponding angles** are equal.

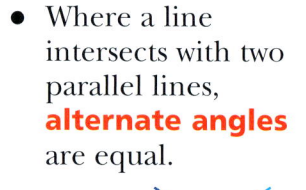

  *Look for the letter F*

- Where a line intersects with two parallel lines, **alternate angles** are equal.

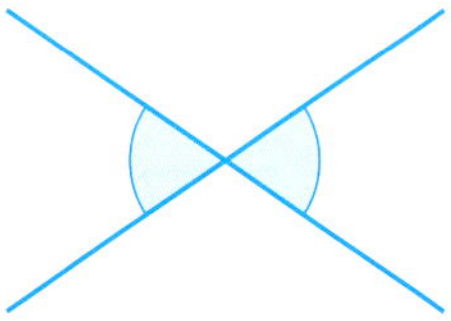

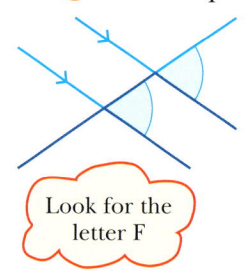

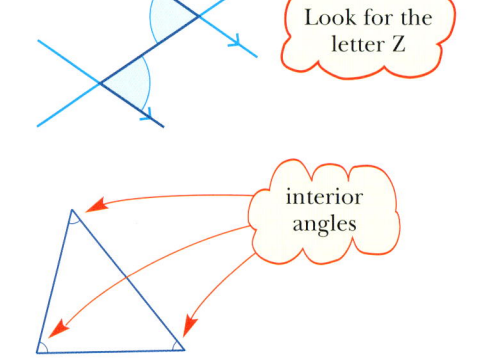

  *Look for the letter Z*

- The angles in a triangle add up to 180°. These are the **interior angles**.

- The angles in a quadrilateral add up to 360°.

## Revision exercise

**1** Find the angles marked with letters in these diagrams.

The diagrams are not drawn accurately.

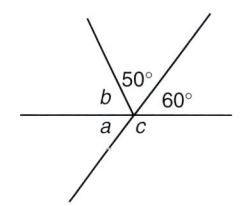

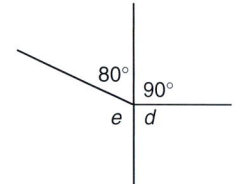

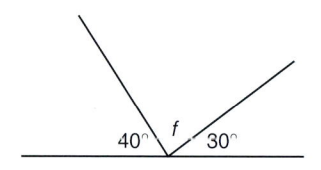

(Diagram 1: angles 50°, 60°, and a, b, c around intersection point)

(Diagram 2: angles 80°, 90°, and d, e around intersection point)

(Diagram 3: angles 40°, 30°, and f on a straight line)

## 3: Triangles and polygons

**Revision exercise**

**2** Find the angles marked with letters in these diagrams.

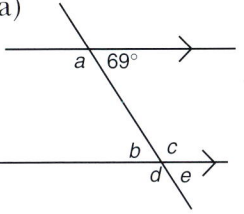

a) 54°, 78°, 85°, 61°, a

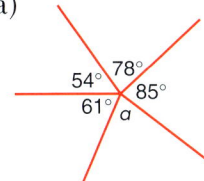

b) 48°, b, 36°

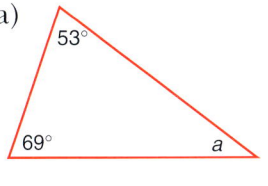

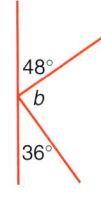

c) 118°, c, 88°, d

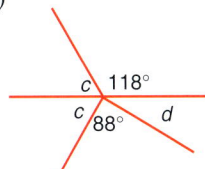

d) 77°, e, f, g

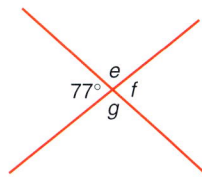

**3** Find the angles marked with letters.

a) a, 69°, b, c, d, e

b) f, g, 65°, h, r, q, s, p, i, j, l, k, n, m, o, 75°

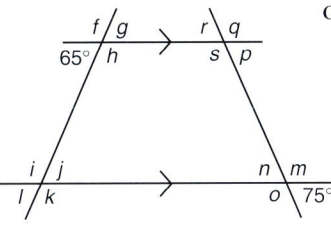

c) 50°, x, t, u, v, w

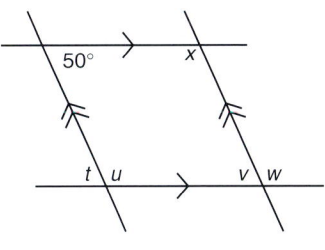

**4** Find the angles marked with letters.

a) 53°, 69°, a

b) 39°, b, c, d, e, 67°, f

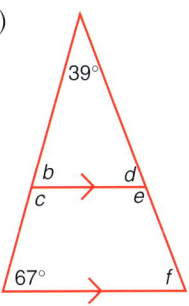

c) 35°, 56°, g, h, 44°, i

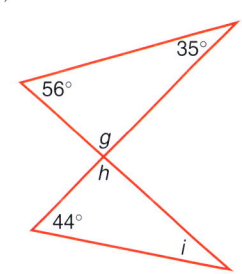

**5** Find the angles marked with letters.

110°, 150°, a, a

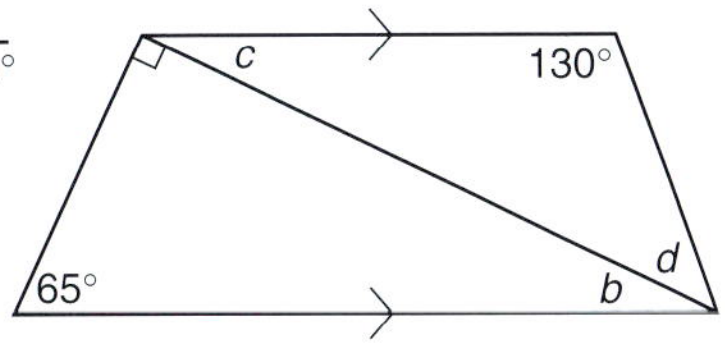

c, 130°, 65°, d, b

g, h, j, f, e, 80°, 60°, i, k

l, 40°

15°, o, n, m, 85°

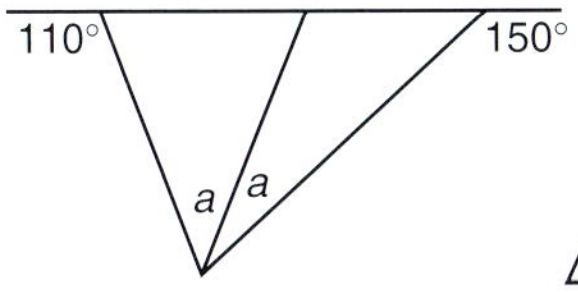

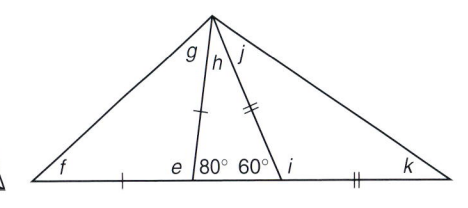

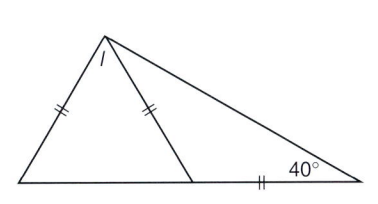

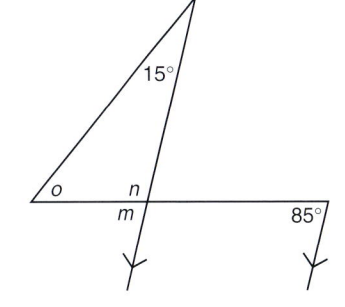

## 3: Triangles and polygons

# Angles and triangles

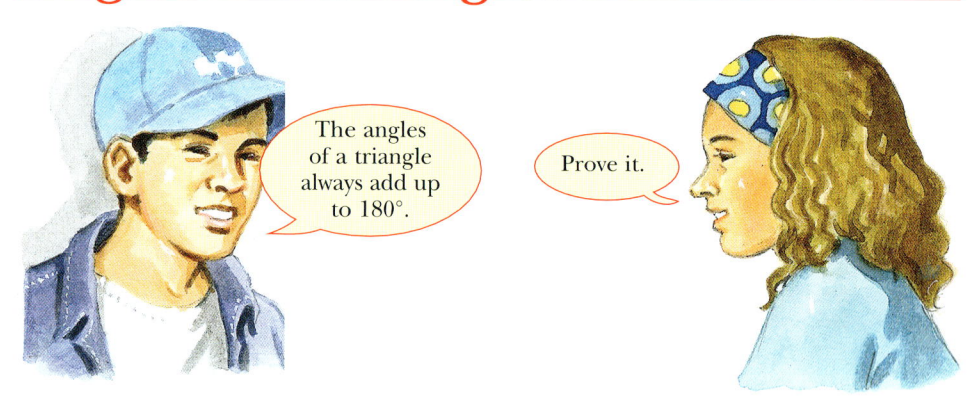

*The angles of a triangle always add up to 180°.*

*Prove it.*

Look at this triangle. One side has been extended.

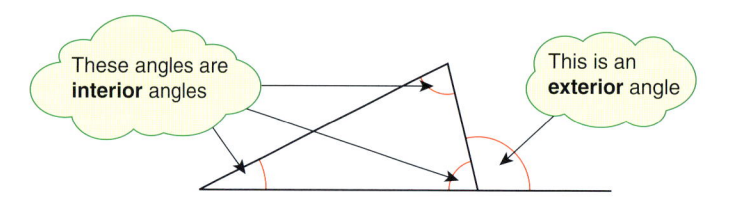

These angles are **interior** angles

This is an **exterior** angle

**?** What do the words interior and exterior mean?

Notice the arrows on two of the lines.

**?** What do they tell you?

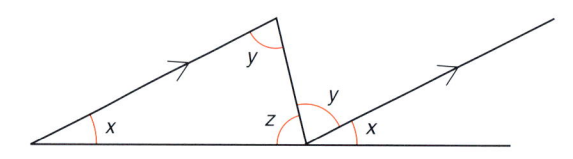

**?** Why are the angles marked *x* the same size?

Why are the angles marked *y* the same size?

The exterior angle of the triangle is *x* + *y*.

> **The exterior angle of the triangle is equal to the sum of the interior opposite angles.**

Look at the straight line which forms the exterior angle.

**?** What do you know about *x* + *y* + *z*?

> **The angles of a triangle always add up to 180°**

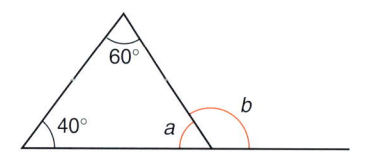

**?** What are the sizes of *a* and *b* in this diagram?

38

## 3: Triangles and polygons

**1** Find the angles marked with letters in these diagrams.

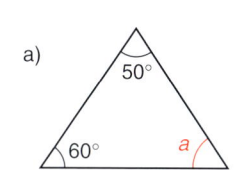

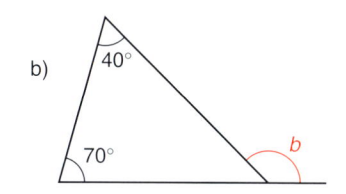

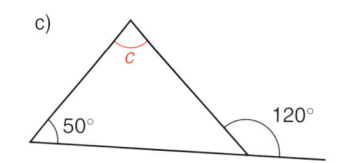

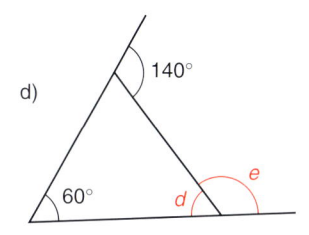

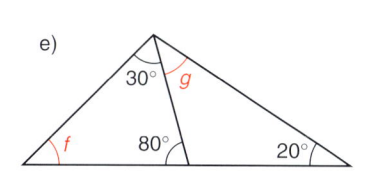

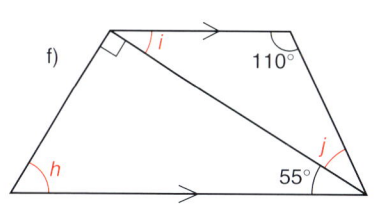

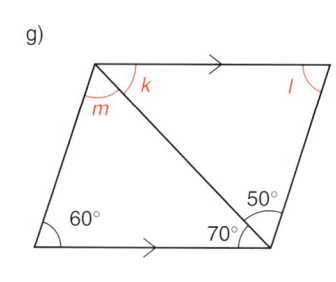

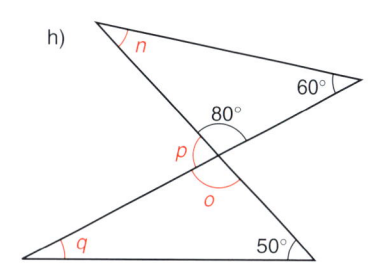

a) triangle with 50°, 60°, a
b) triangle with 40°, 70°, exterior b
c) triangle with c, 50°, exterior 120°
d) triangle with exterior 140°, 60°, d, exterior e
e) triangle with 30°, g, f, 80°, 20°
f) quadrilateral with right angle, i, 110°, h, j, 55° (with parallel marks)
g) parallelogram with k, l, m, 60°, 50°, 70°
h) figure with n, 60°, 80°, p, o, q, 50°

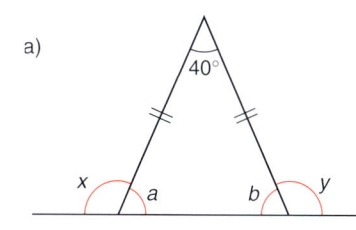

**2**

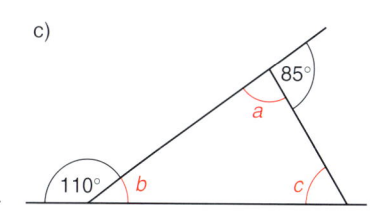

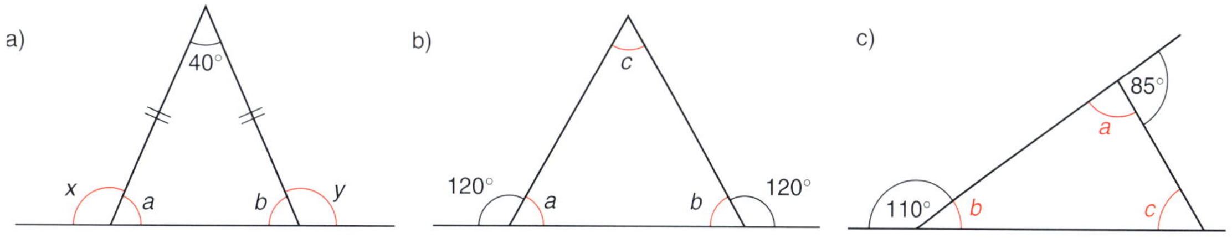

a) isosceles triangle with 40° at top, x, a, b, y on base line
b) triangle with c at top, 120°, a, b, 120° on base line
c) triangle with 85°, a, 110°, b, c

(i) Find the angles marked with letters in the diagrams.

(ii) In each case state the special name which is given to the triangle.

---

Draw any triangle. Colour the three angles in different colours.

Cut off the corners of your triangle. Fit the coloured angles together to make a straight line.

Does it matter how you fit them together? Why is this?

## 3: Triangles and polygons

# Drawing triangles

To draw a triangle accurately, you need to know at least three facts about the length of its sides and the size of its angles. You need a ruler, compasses and a protractor.

## *Drawing a triangle given one side and two angles*

Follow these steps to draw a triangle ABC with side AB = 6 cm long, angle BAC = 62° and angle ABC = 47°.

① Draw and label the side AB (6 cm long).

② Put your protractor at end A of the line and mark off angle BAC. Draw the line.

③ Do the same for angle ABC. Make sure the lines cross.

④ The point where the lines cross is point C.

## *Drawing a triangle given the lengths of all three sides*

Follow these steps to draw a triangle with sides of length 8 cm, 6 cm and 5 cm.

① Draw a line 8 cm long. (You could use any of the sides to start the triangle.)

② Open your compasses to a length of 6 cm. Put the point of the compasses on one end of the 8 cm line. Use the compasses to draw an arc (part of a circle) above the line.

*All points on this arc are 6 cm from A*

③ Now open the compasses to a length of 5 cm. Put the point of the compasses on the other end of the 8 cm line. Draw another arc to cross the first arc. (If you find that you did not draw the first arc far enough, you will have to go back and make it longer.)

*All points on this arc are 5 cm from B*

*This point is 6 cm from A and 5 cm from B.*

④ The point where the arcs cross is the third corner of the triangle. Join this point to both ends of the 8 cm line, and the triangle is complete.

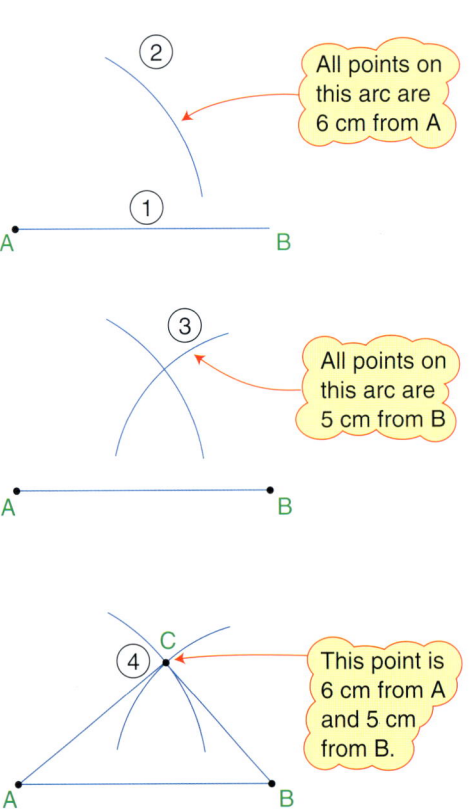

## 3: Triangles and polygons

**1** a) Make an accurate drawing of each of these triangles.

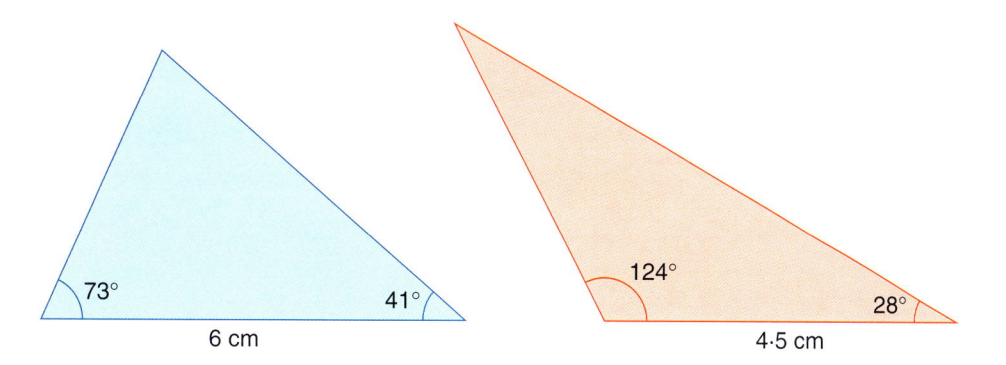

[Triangle 1: angles 73° and 41°, base 6 cm]
[Triangle 2: angles 124° and 28°, base 4·5 cm]

b) Measure the lengths of the other two sides of each triangle. Mark them on your drawings.

**2** Make a rough sketch and then an accurate drawing of each of these triangles. On each accurate drawing, measure and label the lengths of the sides that were not given.

a) Triangle ABC with AB = 7 cm, angle ABC = 46° and angle BAC = 62°.

b) Triangle PQR with PQ = 5 cm, angle PQR = 108° and angle PRQ = 31°.

(Hint: work out angle QPR first.)

**3** Make accurate drawings of triangles with sides as follows. On each drawing, measure and mark on the sizes of all the angles.

a) 3 cm, 5 cm, 6 cm
b) 4.5 cm, 8 cm, 9 cm
c) 4 cm, 5.2 cm, 6.8 cm
d) 7.1 cm, 3.2 cm, 8.4 cm.

**4** a) Try to draw a triangle with sides 7 cm, 4 cm and 2 cm.

b) Explain why it is not possible to draw this triangle.

**5** In each part of this question three lengths are given. Some of these can form the sides of a triangle, others cannot. Without trying to draw the triangles say which can be drawn and which cannot.

a) 3 cm, 6 cm, 8 cm
b) 4 cm, 3 cm, 9 cm
c) 2.5 cm, 4.1 cm, 6.8 cm
d) 5.3 cm, 6.4 cm, 11.1 cm
e) 4.2 cm, 5.7 cm, 9.9 cm

> **3: Triangles and polygons**

# Drawing more triangles

## Drawing a triangle given two sides and the angle between them

Follow these steps to make an accurate drawing of triangle ABC, with side AB 7 cm long, side AC 5 cm long, and angle CAB 54°.

① Draw the line AB 7 cm long.

② Put the protractor at A and mark off the angle CAB. Draw a line.

③ Mark point C on the line 5 cm from A.

④ Join B to C to complete the triangle.

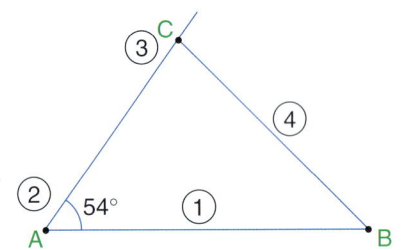

## Drawing a triangle given two sides and an angle not between them

Follow these steps to make an accurate drawing of triangle ABC, with side AB 4 cm long, side AC 7 cm long and angle ABC 110°.

① Draw the line AB 4 cm long. (The first line you draw must be the one with the given angle at one end of it.)

② Use a protractor or angle measurer to mark off an angle of 110° at B, and draw a line. (Make the line longer than you think it needs to be.)

③ Open your compasses to a length of 7 cm. Put the point of the compasses at A. Draw an arc so that it crosses the line you have just drawn from B. Mark point C where the arc crosses the line.

④ Join C to A to complete the triangle.

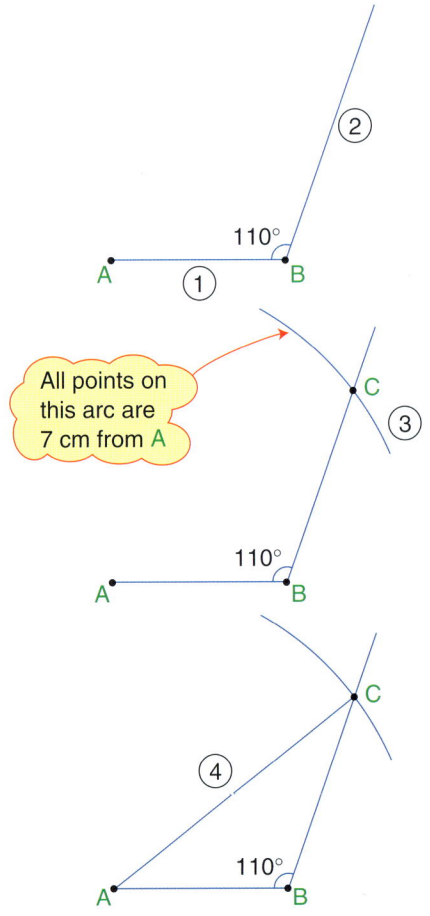

All points on this arc are 7 cm from A

42

# 3: Triangles and polygons

**Exercise**

**1** Make a rough sketch and then an accurate drawing of each of these triangles.

a) Triangle DEF with DE = 8 cm, EF = 3 cm and angle DEF = 57°.
   Measure and mark on your diagram the length of side DF.

b) Triangle XYZ with XY = 4 cm, XZ = 5.6 cm and angle YXZ = 135°.
   Measure and mark on your diagram the length of side YX.

**2** Make a rough sketch and then an accurate drawing of each of these triangles.

a) Triangle FGH with FG = 5 cm, FH = 6 cm and angle FGH = 54°.

   Measure and mark on your diagram the length GH and the angle GFH.

b) Triangle LMN with MN = 3.8 cm, LN = 5.6 cm and angle LMN = 94°.

   Measure and mark on your diagram the length LM and the angle LNM.

**3** You now know how to draw a triangle if you are given

- all three sides but no angles
- two sides and the angle between them
- two sides and an angle not between them
- one side and two angles.

Could you draw a triangle if you were given all three angles but no sides? Explain your answer.

**4** a) Choose a suitable scale and make an accurate scale drawing of each of the triangular sails shown below.

b) Measure on your drawings the angles marked $p$, $q$, $r$ and $s$.

c) Measure on your drawings the sides marked $a$, $b$, $c$ and $d$ and use your scale to work out the real lengths.

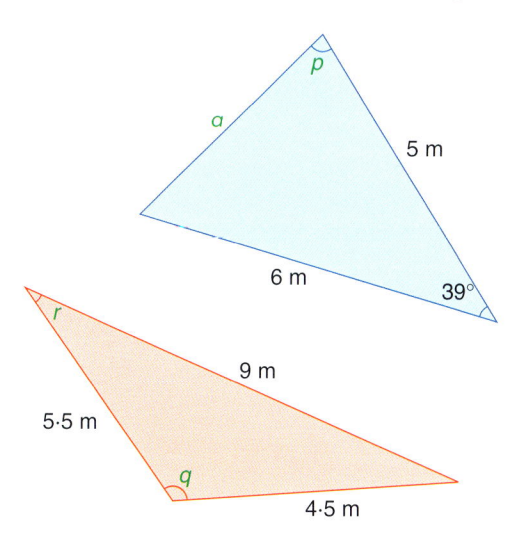

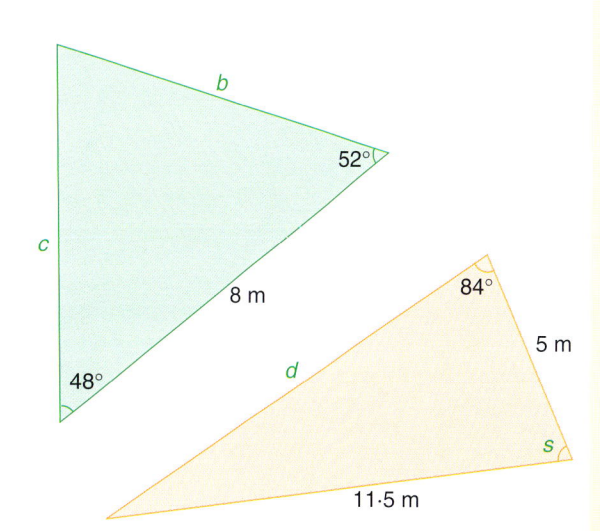

### 3: Triangles and polygons

# Polygons

A **polygon** is a shape with several straight sides.

Some polygons have special names.

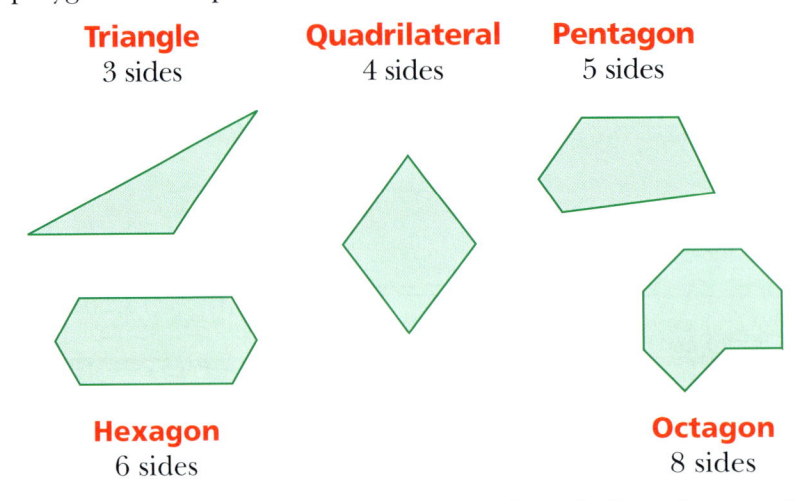

**Triangle**
3 sides

**Quadrilateral**
4 sides

**Pentagon**
5 sides

**Hexagon**
6 sides

**Octagon**
8 sides

Polygons which have all sides the same length and all angles equal are called **regular polygons**.

The shapes shown above are all **irregular polygons**. They do not have equal side lengths and angles.

? *Make sure you can see why each shape is irregular.*

## Interior and exterior angles of a polygon

This diagram shows a pentagon.

The angles shown in red are called **interior angles**.

The angles shown in blue are called **exterior angles**.

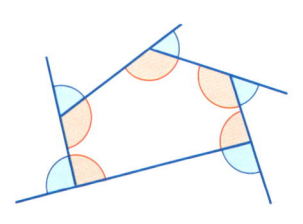

Imagine you are programming a robot to draw a polygon. The exterior angle is the angle through which the robot has to turn after drawing each side.

? *Through how many degrees would the robot have to turn, in total, to draw the whole polygon?*

If a polygon is regular, all the exterior angles are the same size.

For a regular polygon with $n$ sides, each exterior angle = $360° ÷ n$.

? *Why is this?*

## 3: Triangles and polygons

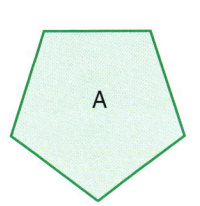

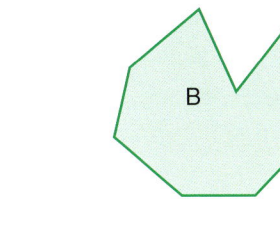

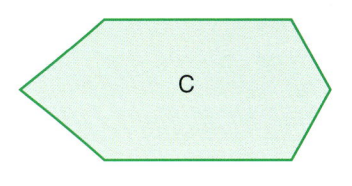

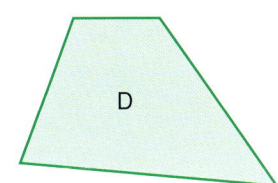

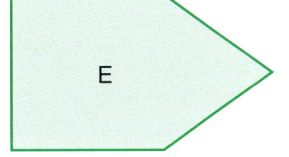

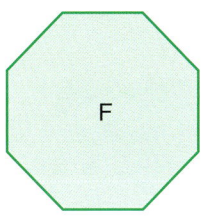

**1** Write down the name of each polygon below and say whether it is regular or irregular.

A    B    C

D    E    F

**2** Find the exterior angle of each of the following polygons.

a) a regular (equilateral) triangle
b) a regular quadrilateral (a square)
c) a regular pentagon
d) a regular hexagon
e) a regular octagon
f) a regular decagon (10 sides)

**3** Here is a decagon made by joining ten equally spaced points on the arc of a circle.

Join each vertex to the centre of the circle and use the diagram to find the size of the interior angles of the decagon.

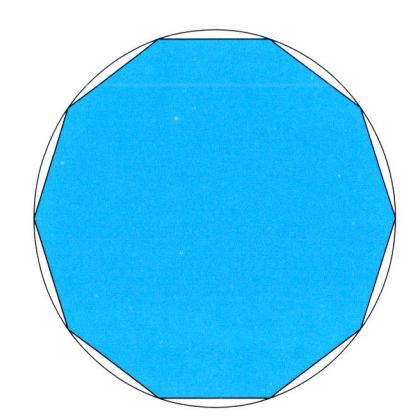

Write a set of instructions for a robot (or write a LOGO program) to draw each of the polygons in question 2. Each instruction should be either FORWARD (plus a distance) or TURN RIGHT/LEFT (plus an angle).

You can choose how big to make each polygon.

# 3: Triangles and polygons

## Angle sum of a polygon

Any polygon can be split into a number of triangles by drawing diagonals.

This pentagon has been split into 3 triangles

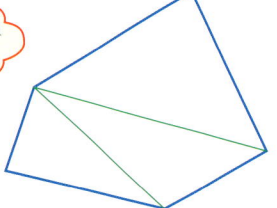

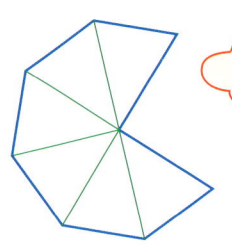

This octagon has been split into 6 triangles

*Into how many triangles could you split a 10-sided polygon?*

*What about a 15-sided polygon?*

*What rule can you use to work out the number of triangles?*

The angles in each triangle add up to 180°, so you can find the sum of the interior angles of the polygon by multiplying the number of triangles by 180°.

*Make sure you can see how this works.*

You can write this as a formula, using $n$ for the number of sides of the polygon:

**Angle sum of a polygon in degrees = $180 \times (n - 2)$**

Using this formula, the angle sum of a pentagon (in degrees)

$= 180 \times (5 - 2)$

$= 180 \times 3$

$= 540$

*Check that this formula works for a triangle ($n = 3$) and a quadrilateral ($n = 4$).*

## Angles of a regular polygon

If a polygon is regular, all the interior angles are the same.

You can find each interior angle by dividing the angle sum of the polygon by the number of angles in the polygon.

The interior angle of a regular pentagon is

$540° \div 5 = 108°$

The number of angles is the same as the number of sides

*What is the relationship between the interior angle and the exterior angle of a polygon?*

# 3: Triangles and polygons

**1** Find the sum of the interior angles of each of these polygons.

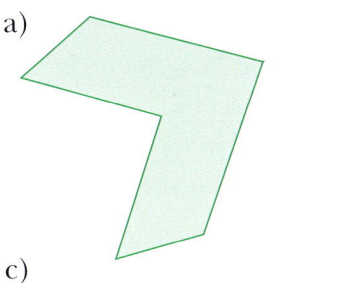

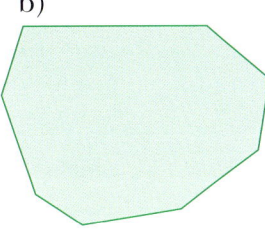

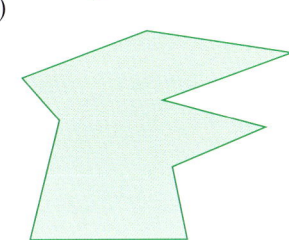

a)

b)

c)

d)

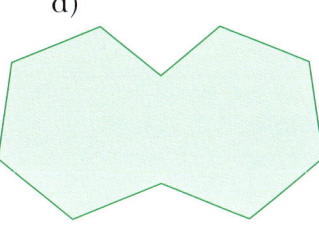

**2** Use your answers to question 1 to work out the interior angle of each of the following regular polygons.

   a) a regular hexagon
   b) a regular octagon
   c) a regular nonagon (9 sides)
   d) a regular decagon (10 sides)

**3** Find the exterior angle of each of the regular polygons in question 2. (Use the rule that for a regular $n$-sided polygon the exterior angle is $360° \div n$.)

**4** a) Use your answers to question 3 to find the interior angle of each regular polygon.

   b) Check your answers with those for question 2.

**5** You have now used two methods for finding the interior angle of a regular polygon: by finding the angle sum first and by finding the exterior angle first.

Which do you find easier, and why?

**6** Find the number of sides in a regular polygon which has an exterior angle of

   a) 30°   b) 40°   c) 60°   d) 90°   e) 36°   f) 72°

The diagram shows part of a tiling pattern made up of regular octagons and squares.
Draw another tiling pattern that uses a regular polygon and one other shape.

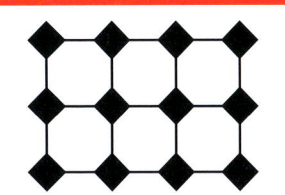

47

## 3: Triangles and polygons

# Finishing off

**Now that you have finished this chapter you should be able to**

★ prove that the exterior angle of a triangle equals the sum of the interior opposite angles

★ prove that the angle sum of a triangle is 180°

★ draw a triangle accurately, given
  – three sides
  – two sides and one angle
  – one side and two angles

★ find the sum of the interior angles of any polygon

★ find the interior and exterior angles of any regular polygon.

Use the questions in the next exercise to check that you understand everything.

**Mixed exercise**

**1** Make accurate drawings of these triangles. Measure and mark on your drawings the lengths and angles marked with letters.

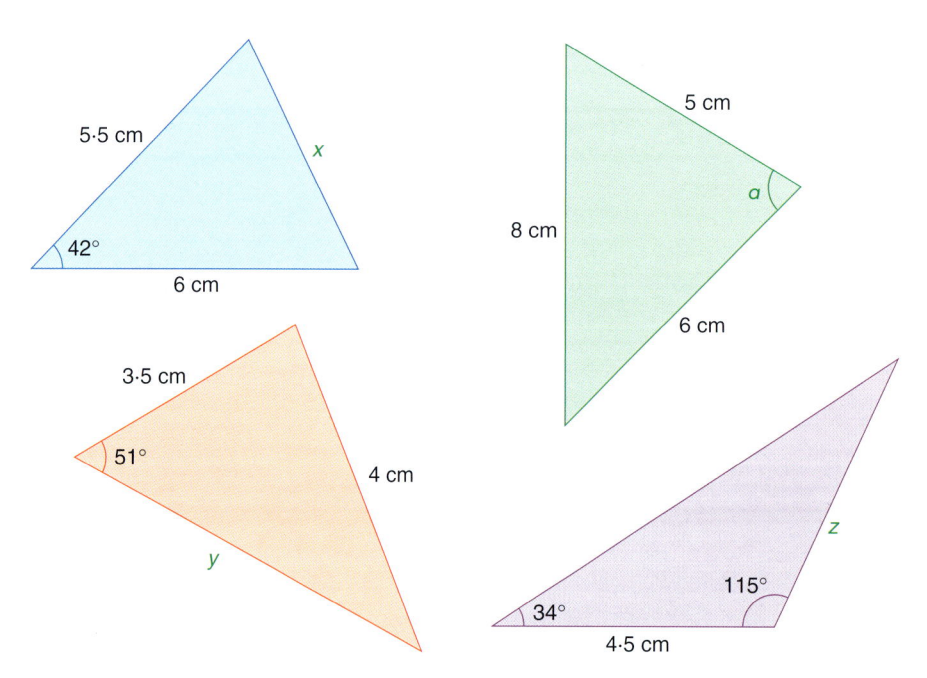

[Triangle 1: sides 5.5 cm and 6 cm with 42° angle between them, angle $x$ marked]

[Triangle 2: sides 5 cm, 8 cm, 6 cm, angle $a$ marked]

[Triangle 3: sides 3.5 cm and 4 cm with 51° angle between them, angle $y$ marked]

[Triangle 4: side 4.5 cm with angles 34° and 115°, angle $z$ marked]

**2** Using only ruler and compasses, construct an equilateral triangle of side 6 cm.

48

## 3: Triangles and polygons

**Mixed exercise**

**3** Find the sum of the interior angles of each of these polygons.

a)

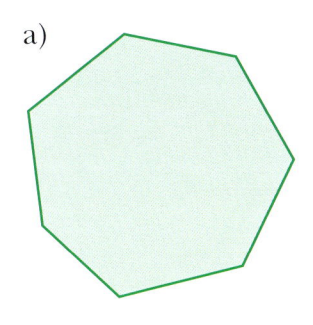

b)

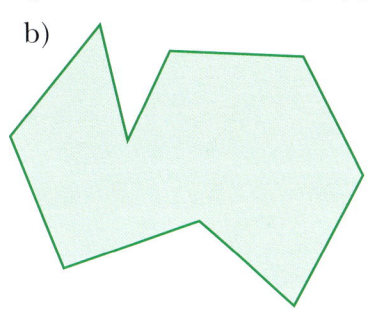

c)

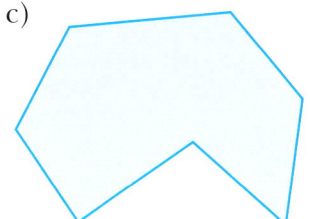

d)

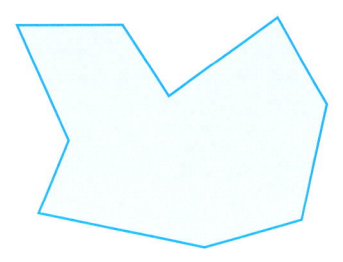

**4** Find the interior angle and exterior angle of

a) a regular hexagon (6 sides)

b) a regular decagon (10 sides)

c) a regular dodecagon (12 sides).

**5** Work out the number of sides of a regular polygon with interior angle

a) 140°   b) 60°   c) 156°

**6** a) Alice, Ben and Carole live at the vertices of a triangle. Alice and Ben live 24 km apart on an east-to-west road. Carole lives north of the road, 20 km from Alice and 20 km from Ben. Using a scale of 1 cm represents 2 km construct this triangle.

b) Don lives 15 km from Alice and 15 km from Ben. Construct the two possible positions of Don's house.

c) How far does Don live from Carole if

(i) Don lives north of the road

(ii) Don lives south of the road?

# Four

# Graphs and co-ordinates

**Before you start this chapter you should**

★ understand what is meant by *x* axis, *y* axis and origin

★ be able to write down the *x* and *y* co-ordinates of a point from a grid

★ be able to plot a point (*x, y*) on a grid.

## Working with co-ordinates

M is the mid-point of AB.

Malik wants to find the co-ordinates of M.

3 is mid-way between 2 and 4.

A (2, 7)
M
B (4, 3)

*What is mid-way between 7 and 3?*

*Do you think that M is at (3, 5)?*

Draw this on squared paper and check whether (3, 5) is the mid-point of AB.

Co-ordinates can also be used to describe a point in three dimensions.

The height above the *x* and *y* axes is measured on a *z* axis.

The seal is at (2, 0, 3).

*Where is the cat?*

*Where is the mouse?*

50

## 4: Graphs and co-ordinates

**Exercise**

**1** Find the mid-point of the line AB in each of the following cases.

  a) A is (1, 4), B is (3, 8)  b) A is (5, 2), B is (9, 8)
  c) A is (6, 0), B is (8, 4)  d) A is (9, 2), B is (3, 6)
  e) A is (10, 5), B is (4, 7)  f) A is (1, 7), B is (3, 9)

**2** Hang-glider R has position (3, 1) and hang-glider T has position (7, 3). Hang-glider S is mid-way between R and T.

Find the co-ordinates of S.

**3** The tops of three chimneys have co-ordinates A (1, 2), C (3, 6) and E (7, 10). Chimney B is mid-way between A and C, and chimney D is mid-way between C and E.

Find the co-ordinates of chimneys B and D.

**4** Five cable cars are travelling up a mountain. Car A is at (3, 8), car C is at (7, 14) and car E is at (11, 20). Car B is mid-way between A and C, and D is mid-way between C and E. Find the co-ordinates of cars B and D.

**5** An aircraft flying at a constant speed on a straight-line path is at A (2, 5). One minute later it is at B (6, 8). After one more minute it is at C.

Find the co-ordinates of C.

**6** Look again at the scaffolding on the previous page.

Give the positions in three dimensions of

  a) the fox     b) the kangaroo   c) the penguin
  d) the elephant  e) the camel     f) the footballer
  g) the blackbird h) the owl       i) the dog.

**7** A stepped pyramid with a square base is shown in the diagram. Each step has a height of 3 units and each step has a width of 1 unit. So A is (1, 1, 6).

Given that B is (12, 0, 0), find the co-ordinates of C, D, E and F.

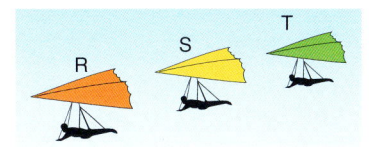

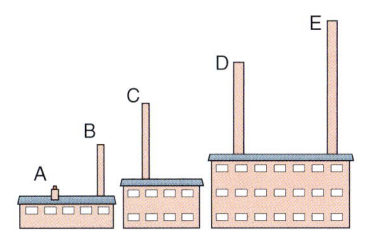

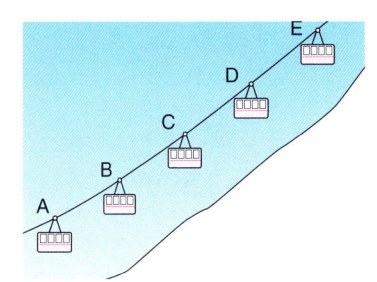

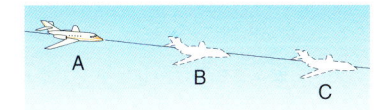

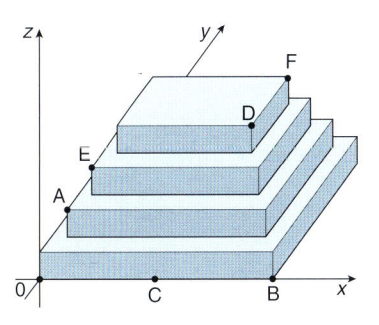

---

Choose one corner of the classroom as the origin.

Take measurements of your classroom and use them to give co-ordinates for the pen on your desk, the top and bottom of the legs on your chair, the whiteboard and the position of the light switch.

# 4: Graphs and co-ordinates

## Using co-ordinates

David and Elaine have found three corners, P, Q and R, of Amelia's house.

They have drawn a grid over the site.

*Amelia's Summer House*
*The house is a parallelogram with a post at each corner.*

Where is the fourth corner? Where do we dig?

M is the mid-point of PR. You can use it to find S.

**?** What are the co-ordinates of S?

To find S you used a fact about parallelograms.

**?** Complete this statement: 'The diagonals of a parallelogram … each other.'

There is another way to find S. From Q to R is 6 East and 2 North. So go to P and then walk 6 East and 2 North.

**?** Where does this place S?

This time you used two more facts about parallelograms.

**?** What are the missing words in this statement?

'Opposite sides of a parallelogram are … and ….'

There are two other places S could be.

**?** Where are they?

These parallelograms are PSQR and PQSR

# 4: Graphs and co-ordinates

**1** Three vertices of a parallelogram are at (1, 1), (–1, –2) and (–3, –1).
Find the possible values of the fourth vertex.

**2** Three vertices of a parallelogram are at (4, 1), (–2, –3) and (–6, –1).
Find the possible values of the fourth vertex.

**3** In the diagram A is (1, 2) and B is (4, 6).
a) What are the co-ordinates of C?
b) Use Pythagoras' rule to find the length AB.

**4** Three vertices of a parallelogram ABCD are A (–2, 2), B (–1, 1) and C (6, 8).
a) Find the co-ordinates of D.
b) Find the lengths AC and BD.
c) What is the special name of parallelogram ABCD?

**5** Three vertices of a parallelogram ABCD are A (3, –2), B (7, 6) and C (–1, 2).
a) Find the co-ordinates of D.
b) Find the lengths AB and BC, leaving your answers in surd form.
c) What is the special name of parallelogram ABCD?

**6** Three vertices of a parallelogram ABCD are A (–3, 2), B (–1, –4) and C (5, –2).
a) Find the co-ordinates of D.
b) Find the lengths AB and BC, leaving your answers in surd form.
c) Find the lengths AC and BD, leaving your answers in surd form.
d) What is the special name of parallelogram ABCD?

**7** Show that A (0, 8), B (17, 1) and C (17, –9) lie on the circumference of a circle centre (5, –4), and find the radius of the circle.

**8** Show that ABC is an isosceles triangle when
a) A is (3, 1), B is (1, 5) and C is (5, 3)
b) A is (–1, –3), B is (14, 17) and C is (6, 21).

---

*Draw a parallelogram of your own on grid paper. Write down the co-ordinates of three of the points. Draw the other two parallelograms using these three points. Exchange your three co-ordinates with a friend and find the missing co-ordinates for their three parallelograms. What shape do you get when you draw all three parallelograms on a single diagram?*

## 4: Graphs and co-ordinates

# Graphs for real life

Ainsley is helping to run the Summer Fayre.

He has designed the programmes on A4 paper and wants to get them photocopied.

**CLONE-a-COPY**
A4 SIZE
£8 plus 3p per copy
A3 SIZE
£8 plus 6p per copy

**AVONFORD LIBRARY PHOTOCOPYING**
A4 5p per sheet
A3 10p per sheet
Please see the Librarian

Ainsley draws a graph to help him decide.

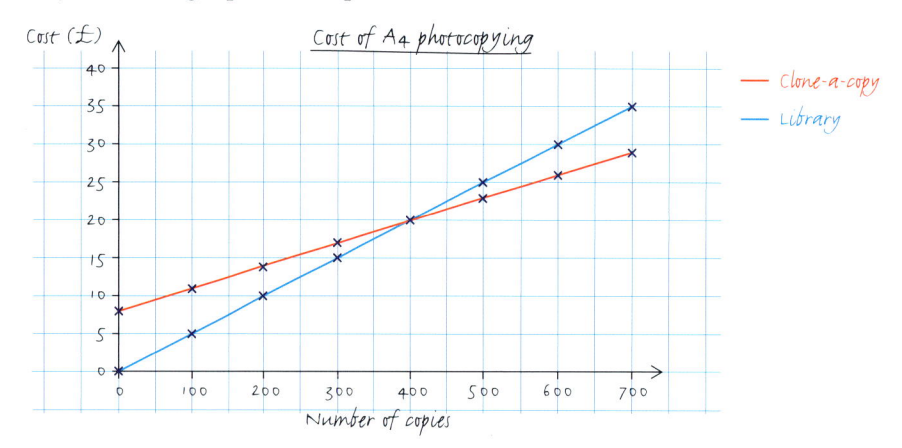

*Cost of A4 photocopying*
— Clone-a-copy
— Library

**?** Which copier should Ainsley choose for 100 programmes?

Which copier should he choose for 700 programmes?

How many programmes would cost the same amount to print at both copiers?

Ainsley also wants some A3 posters.

Draw a similar graph for the posters.

**?** Which copier should Ainsley choose for 100 posters?

Which copier should he choose for 300 posters?

How many posters would cost the same amount to print at both copiers?

**?** What should Ainsley do if he wanted 700 programmes and 50 posters?

How much will they cost him?

# 4: Graphs and co-ordinates

**Exercise**

**1** Avonford Dramatic Society is going to put on a play and they need to hire some stage lights. The company Lights Galore charges £20 per day for a set of lights. Another company, Shadows, hires out an identical set of lights for £130 for up to 7 days plus £10 per day for each additional day.

   a) Draw a graph to compare the two sets of costs.
   b) Which company is cheapest if the lights are required for
      (i) 5 days?    (ii) 8 days?

**2** Mr and Mrs Allan want to hire a car for 7 days. The car-hire company has two different schemes.

   a) Draw a graph to show the cost of hiring a car on each scheme to do up to 1500 miles.
   b) What mileage would cost the same on both schemes?
   c) Mr and Mrs Allan plan to tour around doing about 200 miles per day.

   Which scheme should they choose and what is the price difference?

**CAR-ABOUT** £125 per week plus 6p per mile

**WORRY-FREE** £185 per week NO EXTRAS

**3** Three companies are providing mobile phone facilities with the quarterly charges shown below.

| Company | Quarterly rental | Cost per minute |
|---|---|---|
| Teletalk | £8 | 5p |
| Cheap-Chatter | £30 | 1p |
| Fone-a-friend | free | 10p |

   a) Draw a graph to compare these quarterly costs for up to 600 minutes' use a quarter.
   b) Which company would you recommend for
      (i) Jan, who uses the phone for about 10 hours a quarter?
      (ii) Liz, who uses the phone for about 5 hours a quarter?
      (iii) Pat, who uses the phone for about 2 hours a quarter

**4** Mike is a travelling salesman. He is paid 50p per mile for the first 600 miles each month. After he has travelled 600 miles his expenses are reduced to 40p per mile.

   a) Draw a graph for Mike to use as a ready reckoner for his monthly expenses.
   b) How much is he paid for a monthly mileage of
      (i) 400 miles?    (ii) 800 miles?    (iii) 1200 miles?

## 4: Graphs and co-ordinates

# Number patterns and sequences

Suzie works for a travel company. She has a file containing sheets like this one to help her work out the costs of holidays.

*Hotel del Puerto*

| Nights | Cost (£) |
|--------|----------|
| 1 | 150 |
| 2 | 180 |
| 3 | 210 |
| 4 | 240 |
| 7 | 330 |
| 14 | 540 |
| 21 | 750 |

Look at the first four numbers in the cost column.

They form a **sequence**.

| Number of nights, $n$ | 1 | 2 | 3 | 4 |
|---|---|---|---|---|
| Cost, $C$ (£) | 150 | 180 | 210 | 240 |

*The 1st term in the sequence is 150. The second is 180, and so on*

**?** *How do you go from one term to the next?*

This sequence can be written as a formula:

$$C = 120 + 30n$$

*$30n$ makes $C$ increase in jumps of 30*

**?** *Is the formula right for $n = 7$, 14 and 21?*

**?** *What costs £120? What costs £30?*

*How could you say the formula in everyday English?*

Suzie could have drawn a graph like this. She could read off the cost for any number of nights as shown.

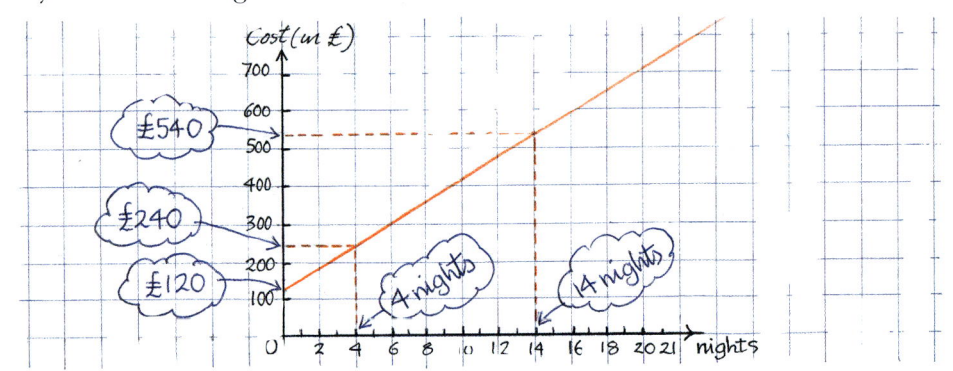

**?** *Which do you think is the most helpful for Suzie: the table, the formula or the graph?*

## 4: Graphs and co-ordinates

**Exercise**

**1** Write down the first 8 terms and the $n$th term of these sequences. Write the term number above each term.

a) 2, 4, 6, 8, …   b) 10, 12, 14, …

c) 1, 3, 5, 7, …   d) 0, 3, 6, 9, …

e) 1, 4, 9, 16, …   f) $1 \times 2$, $2 \times 3$, $3 \times 4$, $4 \times 5$, …

**2** Look at this sequence.

$1, 1 + 3, 1 + 3 + 5, 1 + 3 + 5 + 7, \ldots$

You can see that when you work them out, the first four terms are 1, 4, 9 and 16.

a) Write down the next two terms in the sequence and work out their values.

b) Without working out any more terms, complete these sentences.

*When the first 10 odd numbers are added together, their sum is …*

*When the first n odd numbers are added together, their sum is …*

**3** One week the lottery jackpot is £12 million. Jim has a winning ticket.

a) Calculate how much Jim wins if the number of people with winning tickets (including himself) is

(i) 1  (ii) 2  (iii) 3

(iv) 4  (v) 5  (vi) $n$

b) Jim finds that by mistake he has filled in two identical tickets, so both are winners. How much does he win now for each of the cases in part a)?

c) Which do you think is better, to fill in two identical tickets or two different tickets?

**4** a) Copy and continue this sequence until it has 6 terms:

$1 \times 0 + 1$, $2 \times 1 + 2$,

$3 \times 2 + 3$, $4 \times 3 + 4$, …

b) Write down the 8th term and the $n$th term in this sequence.

c) Work out the terms in a) to make a new sequence and write down the 8th term and the $n$th term in this new sequence.

d) Use algebra to explain why the terms in the first sequence work out to be squares.

---

How many coins do you need if you want to be able to give the exact money for a) any price up to 10p
b) any price up to £1?

**Exact fare please**
**No change given**

c) What about any price up to £10?

# 4: Graphs and co-ordinates

## Finishing off

**Now that you have finished this chapter you should be able to**

★ find the co-ordinates of the mid-point of a line

★ find the position of a point in three dimensions using co-ordinates

★ use geometrical properties of parallelograms to find co-ordinates of special points

★ draw straight line graphs to represent real-life situations

★ work out the $n$th term of a sequence of numbers

★ represent a sequence as a formula.

Use the questions in the next exercise to check that you understand everything.

### Mixed exercise

**1** Find the mid-point of the line AB in each of the following cases.
   a) A is (1, 4), B is (5, 10)
   b) A is (−1, 2), B is (5, 4)
   c) A is (−3, 0), B is (−1, 4)
   d) A is (2, 2), B is (6, 20)
   e) A is (10, 5), B is (−2, −7)
   f) A is (1, 7), B is (−1, −7)

**2** Helicopter R has position (5, 3) and helicopter T has position (13, 9). Helicopter S is mid-way between R and T.

Find the co-ordinates of S.

**3** The tops of three TV aerials have co-ordinates A (−1, −2), C (5, 8) and E (11, 18). B is mid-way between A and C and aerial D is mid-way between C and E.

Find the co-ordinates of B and D.

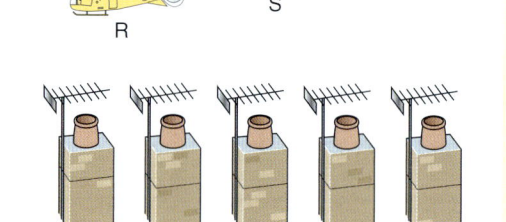

**4** A is (−1, 5), B is (3, 2) and C is (6, 6).
   a) (i) Find the position of D such that ABCD is a parallelogram.
      (ii) Calculate the lengths of AB and AD.
      (iii) Calculate the lengths of AC and BD, leaving your answers in surd form.
      (iv) Name the special parallelogram ABCD.
   b) (i) Find the position of E and F such that ACBE and ACFB are parallelograms.
      (ii) Calculate the lengths of DE and DF.
      (iii) Calculate the length of EF, leaving your answer in surd form.
      (iv) Give two properties of the triangle DEF.

## 4: Graphs and co-ordinates

**Mixed exercise**

**5** One day the exchange rate was £1 was worth 2.5 German marks. On the same day £3 was worth 10 Dutch guilders.

Draw a graph to show both exchange rates from £0 to £200.

a) (i) Find how many German marks were worth £150.
(ii) Find how many Dutch guilders were worth £150.
b) (i) How many German marks were worth 200 Dutch guilders?
(ii) How many Dutch guilders were worth 300 German marks?

**6** Work out the next two terms and the $n$th term of these sequences.

a) 4, 7, 10, 13, ...    b) 3, 7, 11, 15, ...

c) 9, 8, 7, 6, ...    d) 17, 15, 13, 11, ...

**7** What is the next pattern in this sequence?

Write down the number of dots in each of the patterns.

What do you notice?

**8** A crocus bulb splits into 3 each year.

Write down the next 5 terms in the sequence of the number of new bulbs each year.

1, 3, 9, . . .

**9** A small booklet is made by folding a sheet of A4 paper in half.

a) Now another sheet is folded and the two sheets are stapled like this:

How many pages does this booklet have?

b) Write down a sequence of the number of pages in a booklet made in this way.

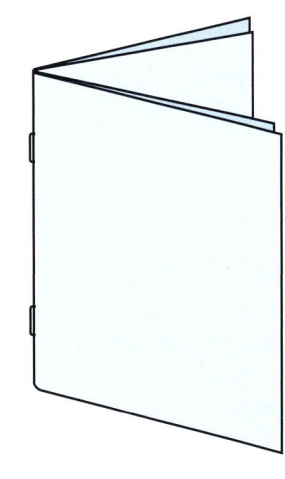

# Five

# Simultaneous equations

**Before you start this chapter you should be able to**

★ solve simple equations.

## Using simultaneous equations

Lisa and George enter a fishing contest.

Their catches are recorded as shown.

The roach are all of a similar size – they are from the same shoal. The same is true of the perch.

After the contest the fish are put back and they swim away.

Later on, Lisa and George try to work out from their results the weights of a typical roach and perch.

**AVONFORD ANGLERS**

**ANNUAL CONTEST RESULTS**

| Name | Catch | Weight |
|---|---|---|
| Lisa Gray | 5 roach 2 perch | 1600g |
| George Gray | 3 roach 2 perch | 1200g |

*How can they work this out?*

*Puzzle it out, then discuss how you did it.*

One approach is to write and solve a pair of **simultaneous equations**, as shown below.

Using $r$ and $p$ for the weights of a typical roach and perch, you can write

| | | |
|---|---|---|
| For Lisa's catch | $5r + 2p = 1600$ | ① |
| For George's catch | $3r + 2p = 1200$ | ② |
| Subtract ② from ① | $2r = 400$ | |
| Divide by 2 | $r = 200$ | |
| Substitute $r = 200$ in ① | $5 \times 200 + 2p = 1600$ | |
| | $1000 + 2p = 1600$ | |
| Subtract 1000 | $2p = 600$ | |
| Divide by 2 | $p = 300$ | |

*① and ② are the simultaneous equations*

*Subtracting has got rid of the term in $p$. You now have an equation just in $r$, which you can solve*

*Once you have found $r$, you can find $p$ by substitution*

Check: substitute for $r$ and $p$ in the left-hand side of ②:

$3 \times 200 + 2 \times 300 = 1200$ ✓

> The solution is $r = 200$, $p = 300$: a roach weighs about 200 g and a perch about 300 g.

*Why is equation ② used for the check, rather than equation ①?*

60

## 5: Simultaneous equations

**Exercise**

**1** Write down this pair of equations.  $2x + y = 15$  ①
$x + y = 8$  ②

Follow these steps to solve them.

a) Subtract ② from ① and so find $x$.

b) Substitute your answer for $x$ in ① to find $y$.

c) Check that your answers fit equation ②.

**2** Solve these pairs of simultaneous equations. (In some of them it is easier to subtract the first equation from the second.)

a) $2x + y = 17$
$x + y = 13$

b) $2x + y = 21$
$x + y = 11$

c) $x + y = 5$
$3x + y = 17$

d) $x + 4y = 10$
$2x + 4y = 12$

e) $2x + 3y = 19$
$x + 3y = 14$

f) $3x + 5y = 23$
$x + 5y = 11$

g) $5p + 7q = 40$
$2p + 7q = 37$

h) $6a + 5b = 1600$
$11a + 5b = 2100$

i) $4c + 19d = 49$
$11c + 19d = 35$

**3** Solve the following equations. (Remember when you subtract something from itself the answer is always zero. For example, $-y - (-y) = 0$.)

a) $5x - y = 32$
$x - y = 4$

b) $4x - 3y = 11$
$2x - 3y = 5$

c) $7x - 2y = 41$
$3x - 2y = 5$

d) $3x - y = 3$
$x - y = 1$

e) $3x - 4y = 10$
$5x - 4y = 14$

f) $2x - 2y = 9$
$4x - 2y = 19$

**4** Ella sends Sally out to buy 5 packs of white paper and 2 packs of blue paper. The bill comes to £15. The next day Ella sends Sally for 7 packs of white and 2 packs of blue. This time the bill is £19.

What is the price of each type of paper?

**5** Use the information below to form two equations.
Solve them to find the cost of 1 orange and the cost of 1 apple.

£1·30    £1·50

### 5: Simultaneous equations

# More simultaneous equations

Look at this pair of equations.
$7p + 3q = 76$ ①
$p - 3q = 4$ ②

**?** *What happens when you subtract ② from ①?*

In this case, subtracting one equation from the other does not **eliminate** $q$. Instead you have to add the equations.

$7p + 3q = 76$ ①
$p - 3q = 4$ ②
Add $\quad 8p = 80$ ← *q is now eliminated*
Divide by 8 $\quad p = 10$
Substitute for $p$ in ① $\quad 70 + 3q = 76$
Subtract 70 $\quad 3q = 6$
Divide by 3 $\quad q = 2$

**The solution is $p = 10, q = 2$.**

**?** *Check that this solution fits equation ②.*

In many cases, neither adding nor subtracting the equations eliminates an unknown.

For example, look at these.
$5a + 6b = 124$ ①
$7a + 2b = 148$ ②

**?** *Try adding and subtracting to eliminate an unknown.*

In this case, you need to spot that multiplying ② by 3 gives you $6b$ in two equations.

② × 3 $\quad 21a + 6b = 444$ ③ ← *Remember to multiply both sides by 3*
Subtract ① from ③ $\quad 5a + 6b = 124$
$\quad 16a = 320$
Divide by 16 $\quad a = 20$

*Although this equation is labelled ③ actually it is just equation ② written in a different way*

**?** *Use this value of a to find b. Check your solution.*

*Why is it better to subtract ① from ③ rather than the other way round?*

*Describe how you would solve*

$2x + y = 17$
$3x - 2y = 8$

# 5: Simultaneous equations

**1** Solve these equations.

a) $8x + 3y = 86$
$4x - 3y = 34$

b) $x + y = 21$
$2x - y = 33$

c) $5x + 4y = 60$
$8x - 4y = 44$

d) $14y - 3x = 43$
$2y + 3x = 37$

e) $9a - 5b = 68$
$2a + 5b = 9$

f) $3p + 7q = 45$
$3p - 7q = 24$

**2** Write down these equations. $\quad 3x + 4y = 22 \quad ①$
$5x - y = 52 \quad ②$

a) Multiply ② by 4. Call the 'new' equation ③.

b) Add ③ to ①.

c) Find $x$ and substitute in ① to find $y$.

d) Check your solution in ②.

**3** Solve these equations. Don't forget you can use your calculator to help with the arithmetic.

a) $2x + 3y = 73$
$11x - y = 209$

b) $6x + 5y = 8$
$4x + y = 3$

c) $3x - 2y = 20$
$8x - y = 75$

d) $2x + y = 12$
$5x - 4y = 17$

e) $2x - 3y = 7$
$4x - 7y = 13$

f) $2x + 3y = 15$
$4x + 7y = 34$

g) $5x + 2y = 47$
$7x - 8y = 28$

h) $6x - 5y = 50$
$x + 7y = 24$

i) $12x - 4y = 5$
$4x - 8y = 4$

**4** Melissa is buying doughnuts for her family. Most of her family prefer jam ones. She has worked out these possible combinations.

9 jam + 1 ring ⟶ £1·74
7 jam + 4 ring ⟶ £1·74

Work out the price of each type of doughnut.

How many jam doughnuts could she buy for £2?

**5** Carl and his father share the same birthday. Carl's father was 32 when Carl was born.

Today when Carl adds his father's age to his own age the answer is 100 exactly.

How old is Carl and how old is his father?

63

## 5: Simultaneous equations

# Multiplying both equations

**?** How would you solve this pair of simultaneous equations?

$3x + 5y = 19$ ①

$7x - 4y = 60$ ②

This is how Karen solves them.

① × 7:  $21x + 35y = 133$
② × 3:  $21x - 12y = 180$
Subtract:  $47y = -47$
Divide by 47:  $y = -1$

*Remember that $35y - (-12y) = 35y + 12y$*

Substitute in ①:  $3x + 5 \times (-1) = 19$
                   $3x - 5 = 19$
                   $3x = 24$
Add 5
Divide by 3:  $x = 8$

Check in ②:
$7x - 4y = 7 \times 8 - 4 \times (-1)$
$= 56 + 4$
$= 60$ ✓

and so $x = 8, y = -1$

This is how Saleen solves them.

① × 4:  $12x + 20y = 76$
② × 5:  $35x - 20y = 300$
Add:    $47x = 376$  so $x = 8$

Substitute into ①:  $24 + 5y = 19$
                     $5y = -5$
                     so $y = -1$

**?** Do you think one method is easier than the other?

Can you find a method for which you only have to multiply one of the equations?

**?** Instead of subtracting, some people find it easier to change all the signs in the bottom equation. Then they can add it to the top one.

Why does this work?

*If subtracting's bad
Change the sign
On the bottom line
Then add!*

# 5: Simultaneous equations

**1** Write down this pair of equations.

$$3x + 5y = 290 \quad ①$$
$$5x - 2y = 70 \quad ②$$

Follow these steps to solve them.

a) Multiply equation ① by 2.

b) Multiply equation ② by 5.

c) Add your two new equations.

d) Find $x$ and $y$ and check your answers.

**2** Solve each pair of equations.

a) $2x + 3y = 24$
$3x - 2y = 23$

b) $3x + 5y = 24$
$5x + 4y = 40$

c) $3x - 2y = 26$
$4x - 5y = 30$

d) $7x - 6y = 41$
$2x + 5y = 5$

e) $2x + 7y = 35$
$5x + 3y = 15$

f) $10x - 4y = 1$
$16x - 3y = 5$

g) $6x - 5y = 19$
$9x - 2y = 67$

h) $9x + 6y = 4$
$6x + 9y = 1$

**3** Bavinder makes desserts to sell in the family shop. She makes them in big trays and cuts them into portions. A tray of barfi sells for a total of £18 and a tray of halwa sells for £21.

The barfi takes about 45 minutes per tray to prepare and cook, and the halwa takes about 25 minutes per tray.

One morning Bavinder cooks for 5 hours. All her desserts are sold, and the takings from them are £153.

How many trays of barfi and how many trays of halwa did Bavinder make?

**4** David and Sian work part-time at Pizza Palace. After 11 p.m. they are paid at a higher rate.

Last week David was paid £149 for 34 hours' work at basic rate and 5 hours at the higher rate.

Sian was paid £229 for 50 hours at the basic rate and 9 hours at the higher rate.

What is the basic hourly rate and the rate after 11 p.m.?

**5** Jess is a florist. She is preparing the flowers for a wedding.

For the bride's bouquet Jess uses 10 roses and 6 carnations. For each bridesmaid's posy she uses 4 roses and 5 carnations.

The flowers cost her £12 for the bouquet and £6.10 for each posy.

What is the cost per bloom of roses and of carnations?

## 5: Simultaneous equations

# Other methods of solution

### Using graphs

Each Saturday Rob and Ed go to the cinema with friends. Some of them are under 15 and so they get cheaper tickets. Last week only one friend went with them. This equation shows how the group was made up, and the total cost.

$$2x + y = 10 \quad ①$$

*What was the total cost? What do x and y represent?*

*Write down some possible values of x and y.*

The equation for the previous week, when there were more people is

$$4x + 3y = 24 \quad ②$$

The graph of each of these equations is a straight line. You can see that the lines cross at the point (3, 4).

At this point the $x$ and $y$ values satisfy both equations, so the solution to the problem is $x = 3$, $y = 4$.

In other words the cost of tickets is £3 for the under-15s and £4 for the others.

All points on this line have $x$ and $y$ values that satisfy the equation $2x + y = 10$

| x | 0 | 5 |
|---|---|---|
| y | 10 | 0 |

$4x + 3y = 24$

| x | 0 | 6 |
|---|---|---|
| y | 8 | 0 |

All points on this line have $x$ and $y$ values that satisfy the equation $4x + 3y = 24$

*Check that you get the same answer using algebra.*

*Which method do you find easier?*

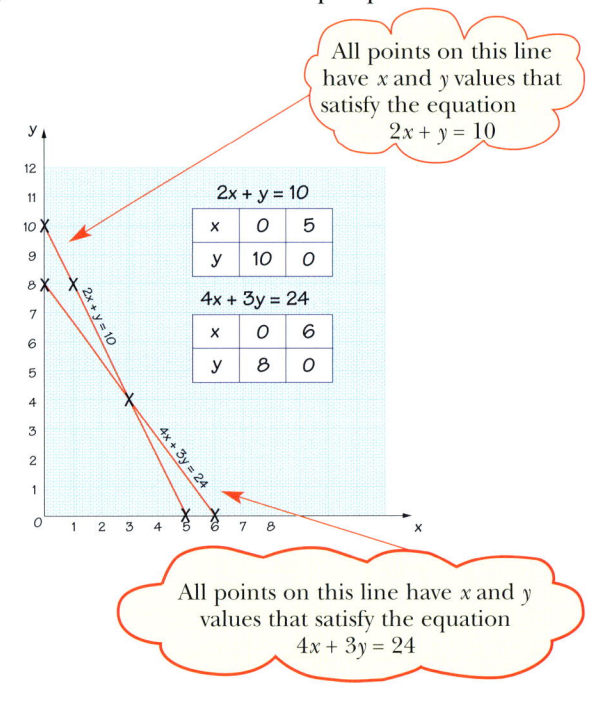

### Using substitution

Another way to solve simultaneous equations is by substitution, as shown below.

$$y = 2x \quad ①$$
$$4x + 3y = 35 \quad ②$$

| | |
|---|---|
| Substitute for $y$ in ② | $4x + 3 \times 2x = 35$ |
| (Tidy up) | $10x = 35$ |
| Divide by 10 | $x = 3.5$ |
| Substitute for $x$ in ① | $y = 7$ |

Substitution is a good method when at least one of the equations is in the form $y = mx + c$

**The solution is $x = 3.5$, $y = 7$.**

*Which method would you use for the equations $x = 3y + 7$ and $2x - y = 24$?*

66

# 5: Simultaneous equations

**Exercise**

**1** Solve each of these pairs of equations using a graph.

For a) to c) draw your *x* and *y* axes from 0 to 10.

a) $x + 2y = 16$
$2x + y = 14$

b) $x + y = 16$
$5x + 3y = 60$

c) $3x + 4y = 24$
$x + y = 7$

For d) to f) draw your *x* axis from 0 to 10 and your *y* axis from −10 to +10.

d) $x + y = 6$
$2x - y = 9$

e) $5x + y = 25$
$x - 2y = 16$

f) $x + 3y = 7$
$4x - y = 2$

**2** Mike collects *x* 200 g jars and *y* 400 g jars to qualify for a free mug.

a) Write down an equation connecting *x* and *y*.

Draw a graph of this equation using axes for *x* from 0 to 12, and *y* from 0 to 10.

b) Mike has bought 8 jars of coffee altogether.

Write down a second equation connecting *x* and *y*.

Draw the graph of this equation on the same axes.

c) Use your graphs to find how many jars of each size Mike has bought.

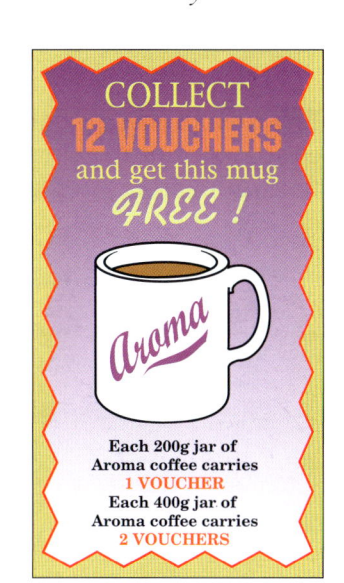

**COLLECT 12 VOUCHERS and get this mug FREE!**

*Aroma*

Each 200g jar of Aroma coffee carries
**1 VOUCHER**
Each 400g jar of Aroma coffee carries
**2 VOUCHERS**

**3** Caroline wants to buy 80 Christmas cards this year. She can do this by buying 8 packets and 2 boxes or 4 packets and 3 boxes.

Write down two equations to show this information.

Solve them by a graphical method to find how many cards are in each packet and each box.

**4** Solve each of these pairs of equations by substitution.

a) $y = 2x + 1$
$5x - 2y = 33$

b) $3x + 2y = 16$
$y = 4x - 3$

c) $y = x - 2$
$y = 8 - 3x$

d) $2y + 3x = 7$
$y = 4 - x$

e) $y = 2x - 3$
$x - y = 2$

f) $y = 7 - x$
$2x + 3y = 15$

**Investigation**

What happens when you try to solve the following pairs of equations by an algebraic method?

a) $4x - 2y = 6$
$10x - 5y = 15$

b) $3x + y = 6$
$3x + y = 12$

Draw graphs for each pair of lines and use them to help you to explain your results.

67

## 5: Simultaneous equations

# Finishing off

**Now that you have finished this chapter you should be able to**

★ solve simultaneous equations using algebra
★ solve simultaneous equations using graphs.

Use the following questions to check that you understand everything.

**Mixed exercise**

1. Solve these equations. In some cases you can add or subtract the equations straight away, but in others you need to multiply first. (Remember that you need to find both unknowns.)

   a) $x + y = 13$
      $x - y = 3$

   b) $x + 5y = 3$
      $x - 4y = 12$

   c) $7p + 4q = 19$
      $2p + q = 5$

   d) $5a - 2b = 7$
      $8a - 4b = 10$

   e) $4x - y = 300$
      $3x - 2y = 0$

   f) $2x + 3y = 16$
      $3x - 2y = 11$

   g) $3s + 2t = 26$
      $2s - 3t = 26$

   h) $2c + 7d = 19$
      $5c - 2d = 28$

   i) $2x + 11y = 52$
      $4x - 11y = 5$

   j) $3x + 20y = 25$
      $4x + 10y = 25$

   k) $4x - 3y = 50$
      $5x - 2y = 80$

   l) $2x + 5y = 72$
      $11x - 2y = 101$

   m) $2x - 3y = 34$
      $11x - y = 1$

   n) $5x + 7y = 6$
      $7x - 3y = 2$

   o) $9x + 10y = 57$
      $3x + 2y = 15$

   p) $4x - y = 36$
      $8x + 2y = 88$

   q) $3x + 5y = 22$
      $2x + 3y = 14$

   r) $x + y = 1$
      $9x + 10y = 11$

   s) $22x + 3y = 116$
      $2x - 5y = 0$

   t) $3x - 8y = -21$
      $7x - 5y = -8$

   u) $8x + 9y = 10$
      $10x - 3y = 3$

2. Mike and Julia have been saving for a while. If they put their savings together they have a total of £1300. Mike has saved £200 more than Julia.

   Write down two simultaneous equations to express this information, and solve them to find out how much each has saved.

3. Miranda asks Rupal to record her two favourite soap operas while she is away on holiday.

   Rupal works out that, using all 480 minutes on the tape, she can record 10 episodes of *Northerners* and 9 of *The Village*, or 8 episodes of *Northerners* and 12 of *The Village*.

   How long are the episodes of each soap?

   Would it be possible to record an equal number of episodes of each soap, and to use up all the tape?

## 5: Simultaneous equations

**Mixed exercise**

**4** The sum of two numbers is 21 and their difference is 5.

Write down two equations to represent this information, and solve them to find the two numbers.

(The sum of $x$ and $y$ is $x + y$ and their difference is $x - y$.)

**5** The Scotts and the Masons buy tickets for a pantomime.

The Scotts buy tickets for 2 adults and 3 children at a cost of £36.

The Masons buy tickets for 4 adults and 2 children at a cost of £48.

a) Work out the cost of an adult's ticket and a child's ticket.

It turns out that 2 of the adults in the Mason party (the grandparents) are entitled to the cheaper tickets because they are over 60.

b) How much refund should the Masons receive?

**6** A company makes two types of medical instrument.

Instrument A takes 4 machine-hours and 7 operator-hours to make.

Instrument B takes 9 machine-hours and 8 operator-hours.

Find how many of each type of instrument the company makes in a week when it devotes 550 machine-hours and 730 operator-hours to making them.

**7** Solve these pairs of equations using graphs.

a) $x + y = 18$
$4x + 3y = 60$

b) $2x + 5y = 40$
$2x - y = 16$

c) $x - y = 4$
$6x + 7y = 63$

**8** Solve each of these pairs of equations by substitution.

a) $y = 3x + 2$
$3x + 2y = 13$

b) $y = 5x + 1$
$y = 7x - 2$

c) $2x - 3y = 9$
$y = 2x - 3$

# Six

# Trigonometry

**Before you start this chapter you should**

★ understand what similar shapes are

★ know that angles round a point add up to 360° and that angles on a straight line add up to 180°

★ know that the angle sum of a triangle is 180°

★ understand and be able to use bearings.

## Introduction to trigonometry

**Trigonometry** is the study of triangles. In this chapter you will learn how to calculate sides and angles in right-angled triangles.

You need a scientific calculator for this chapter.

Triangles A, B and C are all right-angled triangles, and they all have an angle of 30°. This means that they are all enlargements of each other, with different scale factors. They are **similar** triangles.

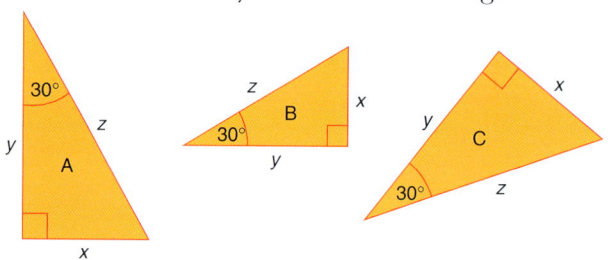

- The sides marked $z$ in the diagram are all opposite to the right-angle. They are the longest sides of each triangle. These sides are called the **hypotenuse**.

- The sides marked $x$ in the diagram are all **opposite** to the angle of 30°.

- The sides marked $y$ in the diagram are all **adjacent** to (next to) the angle of 30°.

For any right-angled triangle with one angle marked, you can label the sides hypotenuse, opposite and adjacent.
(Some people use the abbreviations **hyp**, **opp** and **adj**.)

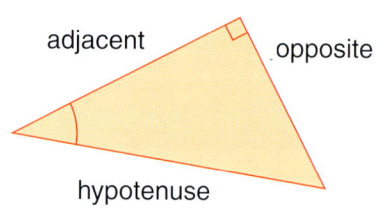

70

# 6: Trigonometry

**1** Copy each of the triangles below and label the hypotenuse, opposite and adjacent sides.

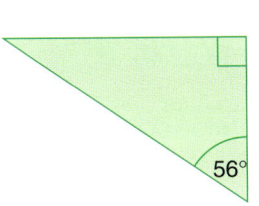

**2** Triangles P, Q and R are similar triangles.

They all have an angle of 40°.

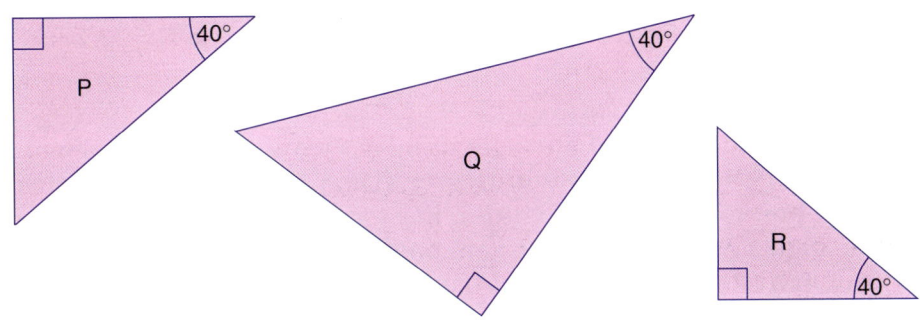

a) Measure the hypotenuse, opposite and adjacent sides of each triangle.

b) Work out the ratio $\dfrac{\text{opposite}}{\text{adjacent}}$ for each triangle.

c) Work out the ratio $\dfrac{\text{opposite}}{\text{hypotenuse}}$ for each triangle.

d) Work out the ratio $\dfrac{\text{adjacent}}{\text{hypotenuse}}$ for each triangle.

e) What do you notice?

**3** Repeat question 2 for triangles X, Y and Z.

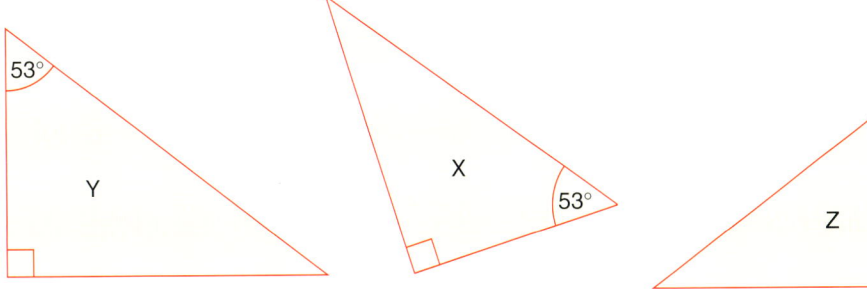

# 6: Trigonometry

## Using tangent (tan)

In the last exercise, you found out that in each of triangles P, Q and R, the ratio $\frac{\text{opposite}}{\text{adjacent}} = 0.84$ (2 d.p.).

This is true for all right-angled triangles with an angle of 40°.

In a similar way, in triangles X, Y, Z and all right-angled triangles with an angle of 53°, the ratio $\frac{\text{opposite}}{\text{adjacent}} = 1.33$ (2 d.p.).

For all right-angled triangles with a particular angle, the ratio $\frac{\text{opposite}}{\text{adjacent}}$ is a fixed number. This number is called the **tangent** (or **tan**) of that angle.

> tan 40° = 0.84
> 
> tan 53° = 1.33

You can find the tan of an angle using a scientific calculator. First, make sure that your calculator is in Degree (Deg) mode. Now check that you know how to use the (tan) button. For some calculators you must enter the angle and then press the (tan) button. For others you must press (tan) first and then the angle, followed by ENTER or EXE.

*Use your calculator to work out tan 40° and tan 53°. Check that you get the answers shown above (although your calculator will give more decimal places).*

*Try to find tan 90° on your calculator. Why doesn't it work?*

### Finding the opposite side using tan

The example below shows how you can use tan to find the length of $x$.

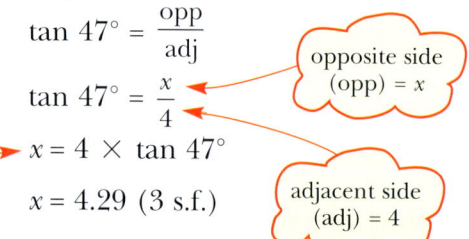

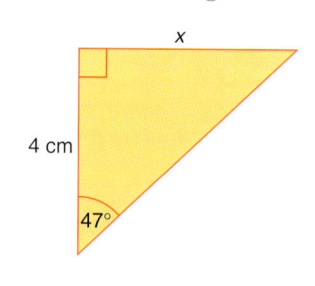

$$\tan 47° = \frac{\text{opp}}{\text{adj}}$$

$$\tan 47° = \frac{x}{4}$$

$$x = 4 \times \tan 47°$$

$$x = 4.29 \text{ (3 s.f.)}$$

- opposite side (opp) = $x$
- adjacent side (adj) = 4

To work this out, first find tan 47° and then multiply by 4

*Did you get the same answer on your calculator? (You should have!)*

# 6: Trigonometry

**Exercise**

**1** Use a calculator to find these tangents. Give your answers to 3 decimal places.

a) tan 25°   b) tan 73°   c) tan 45°

d) tan 37°   e) tan 51°   f) tan 82°

**2** Find the side marked *x* in each of the triangles below.

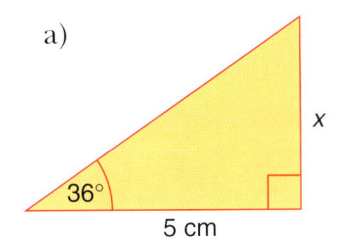

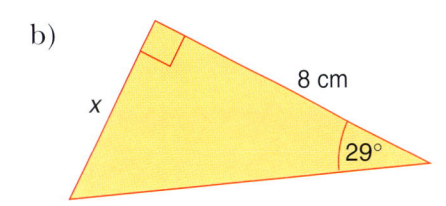

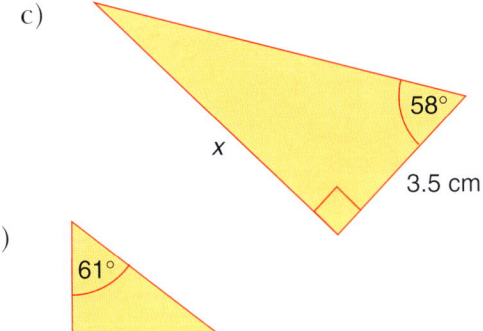

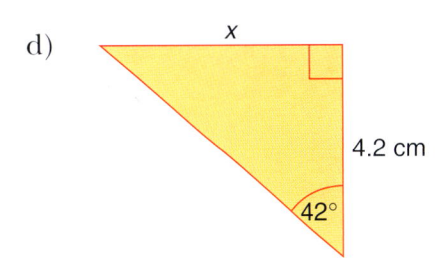

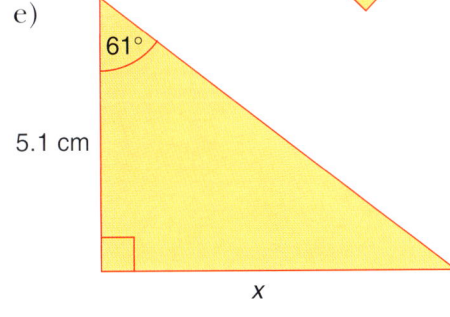

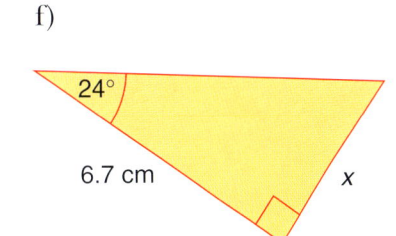

a) [triangle with 36° angle, base 5 cm, opposite side *x*]

b) [triangle with right angle, 8 cm side, 29° angle, side *x*]

c) [triangle with 58° angle, 3.5 cm side, side *x*]

d) [triangle with 42° angle, 4.2 cm side, side *x*]

e) [triangle with 61° angle, 5.1 cm side, side *x*]

f) [triangle with 24° angle, 6.7 cm side, side *x*]

**3** Jenna stands 50 m from the foot of a tall tree. She measures the angle between the ground and a line in the direction of the top of the tree (this is called the **angle of elevation**). She finds that the angle of elevation is 34°. How high is the tree?

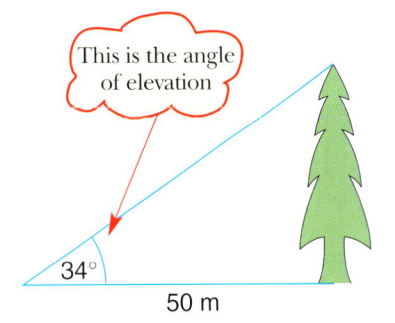

*This is the angle of elevation*

[diagram: tree with 34° angle of elevation at distance 50 m]

Measure the angle of elevation of a tall tree or building, and use it to work out its height.

73

## 6: Trigonometry

# Finding the adjacent side using tan

In the last exercise, all the lengths that you had to work out were opposite sides.

You can use tan to work out an adjacent side, as well.

**Example**

$\tan 72° = \dfrac{\text{opp}}{\text{adj}}$

$\tan 72° = \dfrac{7}{x}$

opposite side (opp) = 7

$x \times \tan 72° = 7$

$x = 7 \div \tan 72°$

adjacent side (adj) = $x$

$x = 2.27$ (3 s.f.)

**The adjacent side is 2.27 cm.**

*Check that you get the same answer on your calculator.*

## Finding an angle using tan

You can also use tan to find the angle in a right-angled triangle if you know the opposite and adjacent sides. The example below shows how.

$\tan x = \dfrac{\text{opp}}{\text{adj}}$

$\tan x = \dfrac{7}{4} = 1.75$

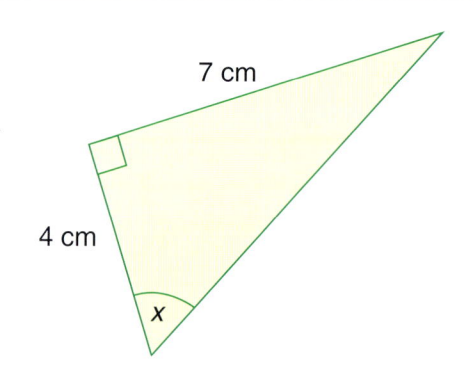

You now need to find out what angle has a tan of 1.75. To do this you need to 'undo' tan. You can do this using a scientific calculator. The key you need may be labelled (tan⁻¹) or (arctan) or you may need to press (INV) followed by (tan). You may have to press the key before or after entering 1.75.

*Use your calculator to work out what angle has a tan of 1.75.*

*You should get the answer 60.3 (correct to 1 decimal place).*

$x = 60.3°$ (1 d.p.)

# 6: Trigonometry

**1** Find the sides marked with letters in each of the triangles below.

a) (8 cm, 64°, side $a$)

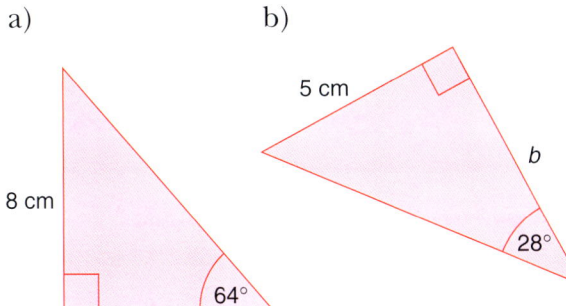

b) (5 cm, 28°, side $b$)

c) (6.2 cm, 47°, side $c$)

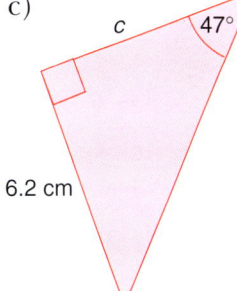

**2** Find the angles marked with letters in each of the triangles below.

a) (5.9 cm, 3.4 cm, angle $d$)

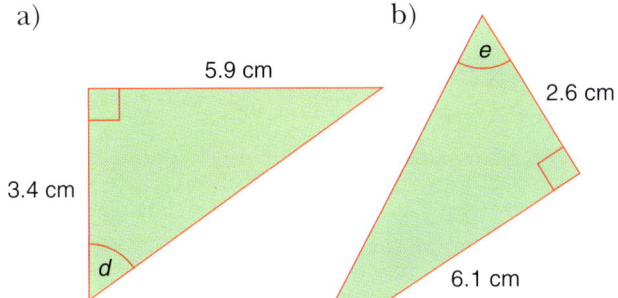

b) (2.6 cm, 6.1 cm, angle $e$)

c) (4.8 cm, 5.2 cm, angle $f$)

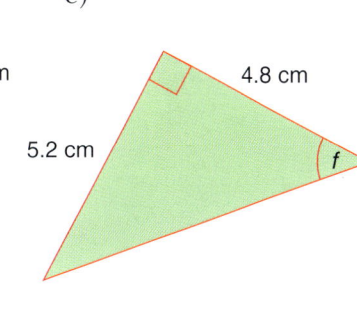

**3** Find the sides and angles marked with letters in each of the triangles below. They are a mixture of the types you have met so far.

a) (3.2 cm, 54°, side $p$)

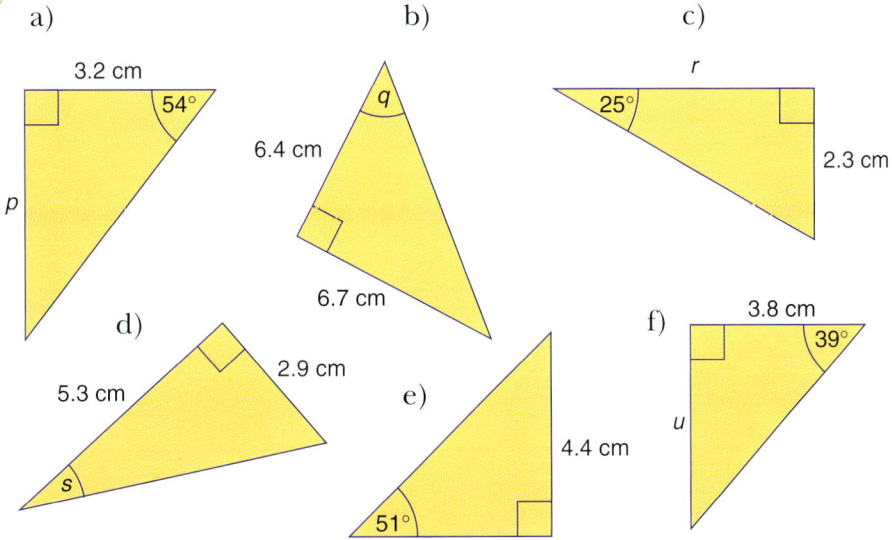

b) (6.4 cm, 6.7 cm, angle $q$)

c) (25°, 2.3 cm, side $r$)

d) (5.3 cm, 2.9 cm, angle $s$)

e) (51°, 4.4 cm, side $t$)

f) (3.8 cm, 39°, side $u$)

# 6: Trigonometry

## Using sine (sin)

In this triangle you can't use tan to find the opposite side, marked *x*, because you don't know the adjacent side.

You need to use another ratio.

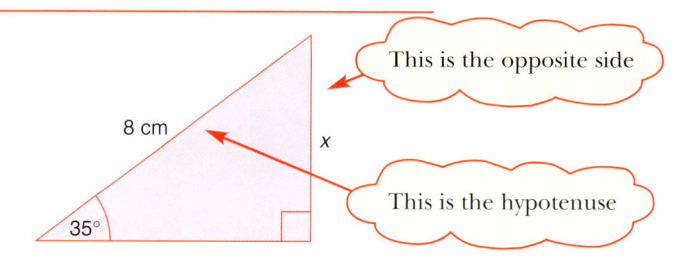

> This is the opposite side
>
> This is the hypotenuse

For a case like this, the ratio $\frac{\text{opposite}}{\text{hypotenuse}}$ is used.

This ratio is called **sine**, or **sin**. You will find a (sin) button on a scientific calculator. It is used in the same way as the (tan) button.

*Use your calculator to find* sin 90°. *Explain the answer you get.*

The examples below show how to use sin to find the opposite side, the hypotenuse or the angle.

### Finding the opposite side using sine

$\sin 57° = \dfrac{\text{opp}}{\text{hyp}}$

$\sin 57° = \dfrac{x}{12}$

$x = 12 \times \sin 57°$

$x = 10.1 \; (3 \text{ s.f.})$

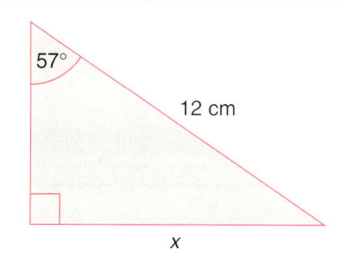

### Finding the hypotenuse using sine

$\sin 29° = \dfrac{\text{opp}}{\text{hyp}}$

$\sin 29° = \dfrac{4.2}{x}$

$x \times \sin 29° = 4.2$

$x = 4.2 \div \sin 29°$

$x = 8.66 \; (3 \text{ s.f.})$

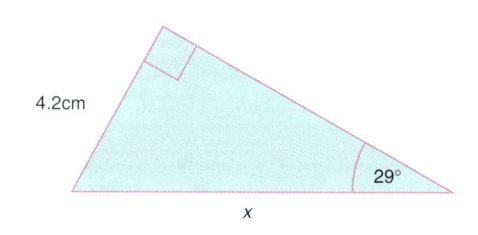

### Finding the angle using sine

> Remember to use $\sin^{-1}$ here

$\sin x = \dfrac{\text{opp}}{\text{hyp}}$

$\sin x = \dfrac{3.8}{6.1} = 0.623$

$x = 38.5° \; (3 \text{ s.f.})$

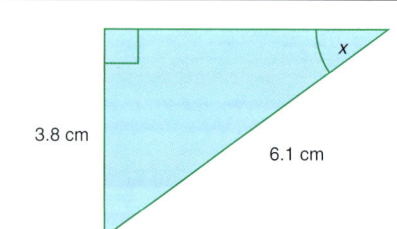

# 6: Trigonometry

**Exercise**

**1** Find the sides marked with letters in each of these triangles.

a) 6.8 cm, 41°, side a

b) 73°, 9.4 cm, side b

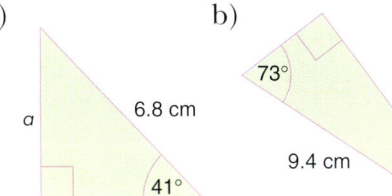

c) 48°, 4.6 cm, side c

d) 62°, 7.1 cm, side d

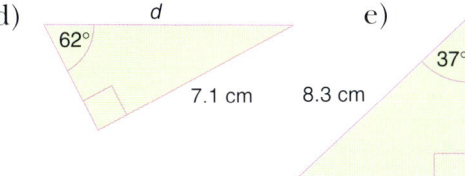

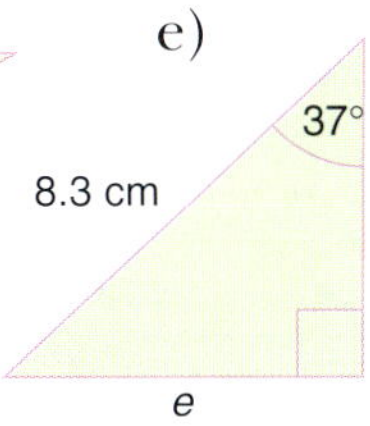

e) 37°, 8.3 cm, side e

f) 5.9 cm, 56°, side f

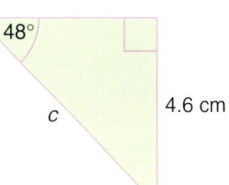

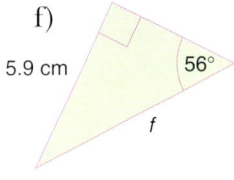

**2** Find the angles marked with letters in each of these triangles.

a) p, 8.2 cm, 6.3 cm

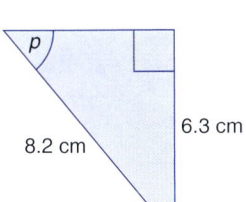

b) 3.7 cm, 6.1 cm, angle q

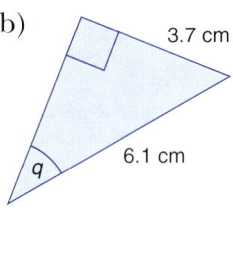

c) r, 12.3 cm, 7.5 cm

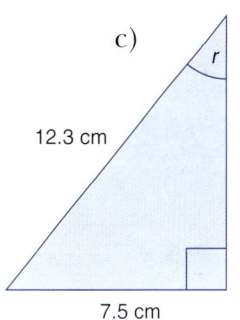

**3** A ladder is 8 m long.

a) For safety reasons, the angle it makes with the ground should not be more than 75°.

What is the highest the ladder can reach?

b) What angle must it make with the ground to just reach a gutter 6 m above the ground?

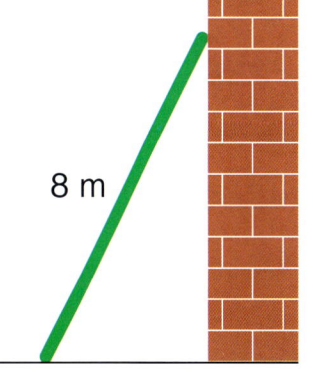

---

Find out what road gradients such as '1 in 10' mean. Use trigonometry to work out the angle of the slope for different gradients.

# 6: Trigonometry

## Using cosine (cos)

You have now met tangent, which links the opposite and adjacent sides, and sine, which links the opposite side and the hypotenuse. A way of linking the adjacent side and the hypotenuse is now needed.

The ratio $\dfrac{\text{adjacent}}{\text{hypotenuse}}$ is called the **cosine** (or **cos**).

It can be used to find the adjacent side, the hypotenuse or an angle in the same way that tangent and sine are used.

*Use your calculator to find cos 90°. Explain the answer you get.*

The examples below show how to use cos to find the opposite side, the hypotenuse or the angle.

### Finding the adjacent side using cos

$\cos 41° = \dfrac{\text{adj}}{\text{hyp}}$

$\cos 41° = \dfrac{x}{8}$

$x = 8 \times \cos 41°$

$x = 6.04$ (3 s.f.)

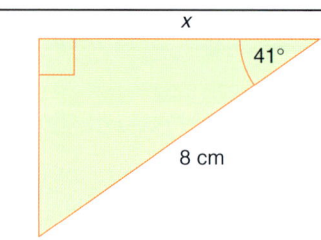

### Finding the hypotenuse using cos

$\cos 68° = \dfrac{\text{adj}}{\text{hyp}}$

$\cos 68° = \dfrac{7.5}{x}$

$x \times \cos 68° = 7.5$

$x = 7.5 \div \cos 68°$

$x = 20.0$ (3 s.f.)

### Finding the angle using cos

Remember to use $\cos^{-1}$ here

$\cos x = \dfrac{\text{adj}}{\text{hyp}}$

$\cos x = \dfrac{4.9}{8.3} = 0.590$

$x = 53.8°$

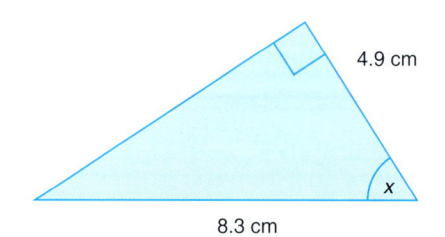

# 6: Trigonometry

**1** Find the sides marked with letters in each of the triangles below.

a) [triangle with right angle, 44° angle, hypotenuse 6.7 cm, side a opposite to 6.7 cm]

b) [triangle with right angle, 31° angle, side 7.2 cm, side b]

c) [triangle with right angle, 48° angle, side 4.6 cm, side c]

d) [triangle with right angle, 62° angle, hypotenuse 11.4 cm, side d]

e) [triangle with right angle, 28° angle, side 9.8 cm, side e]

f) [triangle with right angle, 53° angle, side 3.9 cm, side f]

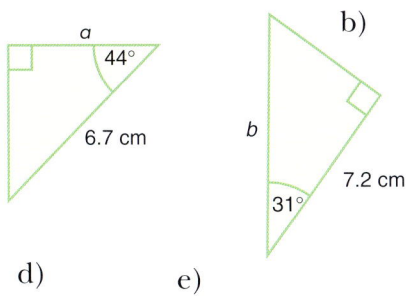

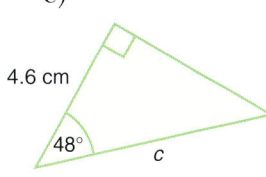

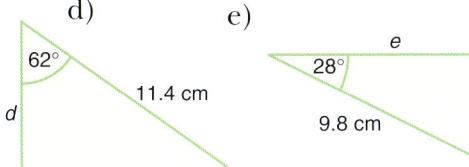

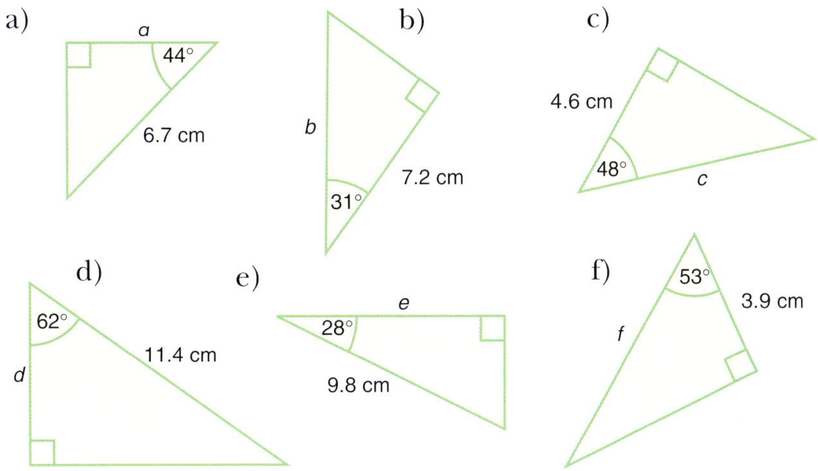

**2** Find the angles marked with letters in each of the triangles below.

a) [right triangle with sides 8.4 cm, 4.5 cm, angle x]

b) [right triangle with sides 6.6 cm, 7.9 cm, angle y]

c) [right triangle with sides 5.2 cm, 10.1 cm, angle z]

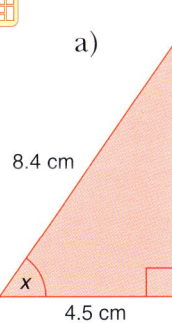

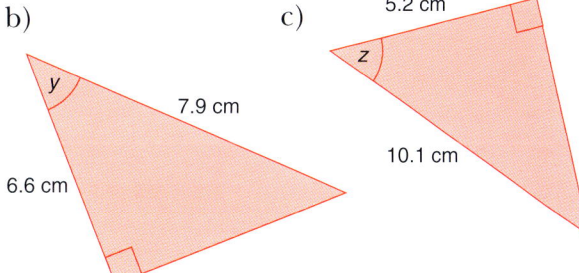

**3** A road slopes at 5° to the horizontal for 2 km measured horizontally, as shown in the diagram below. How long is the road?

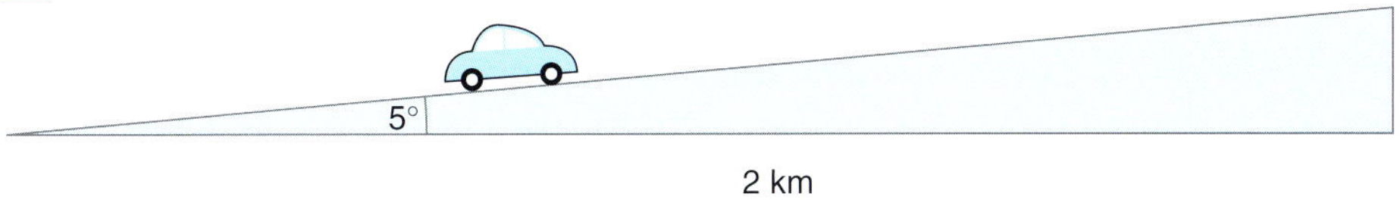

[diagram of car on slope, 5° angle, 2 km horizontal]

79

## 6: Trigonometry

# Using sin, cos and tan

You have now met the three trigonometrical ratios. They are usually written in books or on formula sheets like this:

$$\sin \theta = \frac{\text{opp}}{\text{hyp}}$$

$$\cos \theta = \frac{\text{adj}}{\text{hyp}}$$

$$\tan \theta = \frac{\text{opp}}{\text{adj}}$$

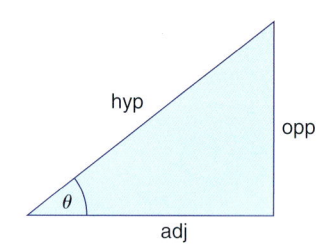

Some people remember these three ratios using the made-up word **SOHCAHTOA**.

**S**in
  **O**pposite
    **H**ypotenuse
      **C**os
        **A**djacent
          **H**ypotenuse
            **T**an
              **O**pposite
                **A**djacent

When you want to find a missing side or angle in a right-angled triangle, the first thing to do is to decide whether to use sin, cos or tan. The example below shows how to do this.

*The side you have been given is the **hypotenuse***

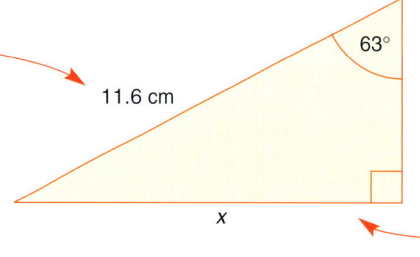

*The side you want to find is the **opposite** side*

11.6 cm, 63°, x

You need the ratio which involves **hypotenuse** and **opposite**. Look at the ratios at the top of the page and you will see that the one you need is sin.

$\sin 63° = \frac{\text{opp}}{\text{hyp}}$

$\sin 63° = \frac{x}{11.6}$

$x = 11.6 \times \sin 63°$

$x = 10.3$ (3 s.f.)

**So the required side is 10.3 cm.**

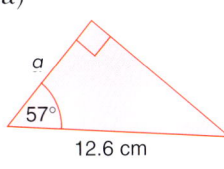

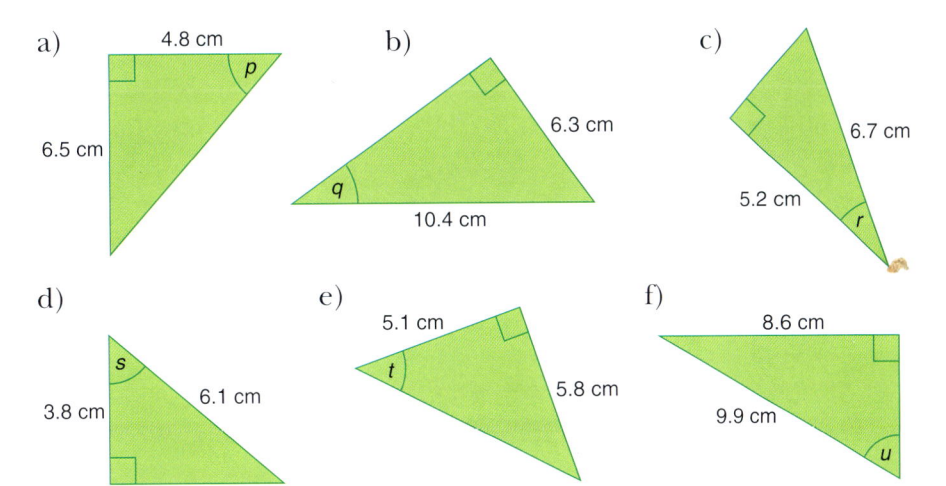

# 6: Trigonometry

**1** Find the sides marked with letters in these triangles.

a) 57°, 12.6 cm, side a

b) 41°, 4.8 cm, side b

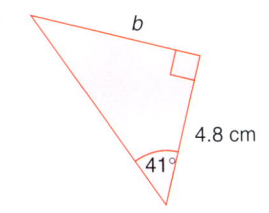

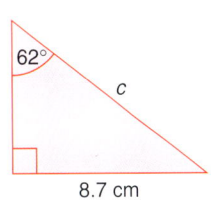

c) 62°, 8.7 cm, side c

d) 36°, 9.7 cm, side d

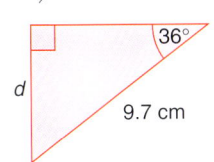

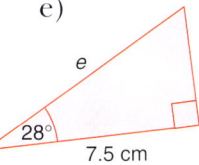

e) 28°, 7.5 cm, side e

f) 55°, 6.9 cm, side f

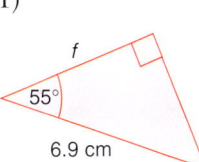

g) 42°, 4.3 cm, side g

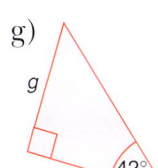

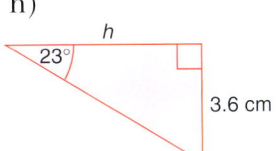

h) 23°, 3.6 cm, side h

i) 34°, 5.1 cm, side i

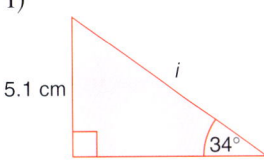

**2** Find the angles marked with letters in these triangles.

a) 4.8 cm, 6.5 cm, angle p

b) 6.3 cm, 10.4 cm, angle q

c) 5.2 cm, 6.7 cm, angle r

d) 3.8 cm, 6.1 cm, angle s

e) 5.1 cm, 5.8 cm, angle t

f) 8.6 cm, 9.9 cm, angle u

# 6: Trigonometry

# Using trigonometry

Trigonometry can be used to solve real-life problems. Always start by drawing a clear diagram showing the right-angled triangle that you are using.

### Example 1

Anna is going to climb a hill which is 1400 m high. She measures the length of the climb on a map and finds that it is 6.5 km.

What is the angle of the slope?

### Solution

*The length of the climb must be the horizontal distance as it has been measured on a map.*

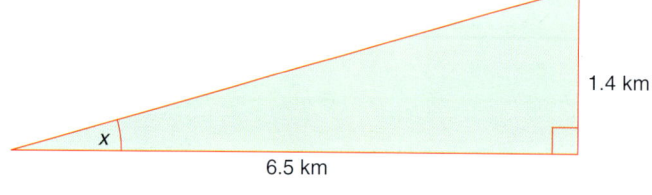

(triangle with x angle, 1.4 km opposite, 6.5 km adjacent)

The two sides marked are the **opposite** and the **adjacent**. So you need to use **tangent** to find the angle.

$$\tan x = \frac{1.4}{6.5} \qquad x = 12.2°$$

**The angle of the slope is 12.2°.**

? *Do you think the slope is really 12.2° all the way up the hill?*

### Example 2

A ship sails 50 km on a bearing of 140°. How far south and how far east is it from its starting point?

### Solution

*This is the path of the ship*

*This is the distance south of the starting point*

*This is the distance east of the starting point*

The side marked *s* is the **adjacent** side, and the side marked 50 km is the **hypotenuse**. So you need to use **cosine** to find *s*.

$$\cos 40° = \frac{\text{adj}}{\text{hyp}} \qquad \cos 40° = \frac{s}{50}$$

$$s = 50 \times \cos 40° = 38.3$$

**The ship is 38.3 km south of its starting point.**

? *Which trig ratio would you use to work out e? Work it out and check you get 32.1 km.*

# 6: Trigonometry

**1** a) An aeroplane flies 300 km on a bearing of 245°. How far south and how far west is it from its starting point?

b) An aeroplane flies 550 km on a bearing of 318°. How far north and how far west is it from its starting point?

**2** A sailor in a small boat can see a cliff in the distance which he knows is 150 m high. The angle of elevation of the top of the cliff is 8°. How far is the boat from the foot of the cliff?

**3** a) A path up a hill slopes at 15°. The path is 3.6 km long. How high is the hill?

b) Another path up the other side of the same hill is 4.3 km long. What is the angle of the slope of this path?

**4** The guy ropes for a tent run from the top of the tent pole, which is 1.5 metres tall, to a point on the ground near the tent. Each rope should ideally be at 45° to the ground.

a) How long should each guy rope be?

b) How far will each rope be from the base of the tent pole?

**5** A short flight of steps, 1.2 m high, is to be replaced by a ramp. The slope of the ramp must not be more than 10°. What is the shortest the ramp could be?

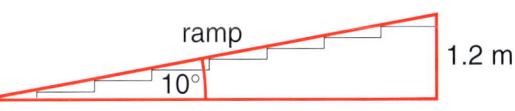

**6** a) A ship starts from point A and sails 200 km on a bearing of 115° until it reaches point B. How far south and how far east is it from point A?

b) The ship then sails a further 150 km from point B on a bearing of 230° until it reaches point C. How far south and how far west is it from point B?

c) How far south and how far east is the ship now from point A?

d) What bearing must the ship sail on to get back to point A directly from point C?

## 6: Trigonometry

# *Finishing off*

**Now that you have finished this chapter you should be able to**

★ find sides and angles in right-angled triangles using sine, cosine and tangent

★ use trigonometry to solve problems involving right-angled triangles.

Use the questions in the next exercise to check that you understand everything.

### Mixed exercise

**1** Find the sides and angles marked with letters in the triangles below.

a) side $a$, with angle 61° and adjacent 4.8 cm

b) angle 48°, hypotenuse side $b$, opposite 6.3 cm

c) top 5.9 cm, side 9.4 cm, angle $c$

d) side 8.1 cm, angle 42°, side $d$

e) 7.3 cm, angle 35°, side $e$

f) angle $f$, 6.9 cm, 4.5 cm

g) 11.7 cm, side $g$, angle 71°

h) 5.1 cm, angle $h$, 10.6 cm

i) angle 56°, 9.3 cm, side $i$

j) 5.8 cm, 8.8 cm, angle $j$

# 6: Trigonometry

**Mixed exercise**

**2** The diagram shows a beam of light from a spotlight in a concert hall 5.4 m high.

  a) Find the radius of the circle of light on the floor.

  b) Find the area of the circle of light.

**3** Simon and Sue set out from a Youth Hostel for a country walk.

They walk 3 km on a bearing of 285°.

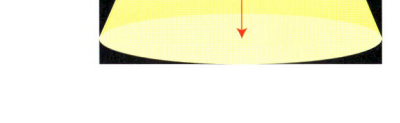

  a) How far north and how far west are they from the Youth Hostel?

  b) They then walk 2 km due north. Find the bearing they need to walk on to get back to the Youth Hostel.

## Investigation

Look at the triangle in this diagram.

$\frac{x}{1} = \cos A$ so $x = \cos A$

$\frac{y}{1} = \sin A$ so $y = \sin A$

The point P has co-ordinates ($\cos A$, $\sin A$).

What happens if angle $A$ is greater than 90°?

Use your calculator to find $\sin A$ and $\cos A$.

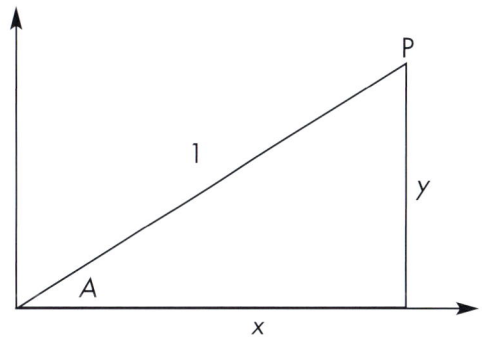

Copy and complete this table for angles at 10° intervals up to 360°.

| angle $A$ | sin $A$ | cos $A$ |
| --- | --- | --- |
| 0° | 0 | 1 |
| 10° | 0.17 | 0.98 |
| 20° | 0.34 | 0.94 |
| 30° | | |

Plot the graph of:

  $\sin A$ against $A$

  $\cos A$ against $A$

What do you notice?

# Seven

# Inequalities

**Before you start this chapter you should be able to**

★ expand brackets

★ solve simple linear equations

★ draw graphs of straight lines from their equations.

## Using inequalities

Look at these notices.

They all describe restrictions on ages, heights and prices.

These restrictions can be written as **inequalities** using special symbols.

< means *is less than*

\> means *is greater than*

≤ means *is less than or equal to*

≥ means *is greater than or equal to*

Read 4 < 7 as *4 is less than 7*

Read 3 > 1 as *3 is greater than 1*

Read $p ≥ 5$ as *p is greater than or equal to 5*

Let $y$ stand for age in years. Then

$y ≥ 15$ describes the age of a person who can go to the film *Wilde*.

$y < 16$ describes the age of a person whose Cinderella ticket will cost £2.50.

**?** *Write inequalities, using the symbols, for the restrictions in the other notices.*

*What other restrictions are commonly shown in notices and signs?*

You have already met the inequality symbols in Book 1, when you were grouping data. In this chapter you will learn about other ways to use them.

**?** *What is the other way of writing 7 > 4?*

# 7: Inequalities

**1** Write these statements as inequalities.

   a) $x$ is greater than 3    b) $x$ is greater than or equal to zero

   c) $x$ is less than 5    d) $x$ is less than or equal to $-2$.

**2** Write these inequalities in words.

   a) $x < 8$    b) $p > 100$    c) $q \geq 100$

   d) $y < 17$    e) $x \leq 20$    f) $b \geq -3$

**3** For each of these, copy the numbers down in the same order, replacing the comma by an inequality sign ($<$ or $>$).

   a) 2, 9    b) 13, 3    c) $-3, -13$

   d) 13, $-3$    e) 3.8, 3.3    f) 0.5, 0.625

**4** Look at each of these signs and write down the restriction shown using the inequality symbols. (Choose a suitable letter, in each case, to stand for the quantity that is being restricted.)

   a) [Speed limit sign: 30]

   b) [Winter Sun! Flights from £79!]

   c) [Old Man Theatre — Party rates for groups of 20 or more]

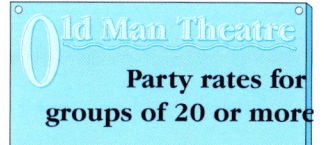

   d) [LIFT MAX LOAD 5 PERSONS]

   e) [Holidays in Spain — 7 nights - less than £250]

   f) [MATCHES — Min. contents 50]

   g) [Narrow Bridge 2·5 m]

   h) [RAINBOW Building Society HIGH INTEREST ACCOUNT Minimum deposit £500]

**5** You can write an inequality in two ways.

For example, $9 > 7$ could be written as $7 < 9$.

Write down each of your inequalities from question 3 in another way.

# 7: Inequalities

## Number lines

It is often helpful to show an inequality on a number line.

This is how you show the inequality $x > 4$.

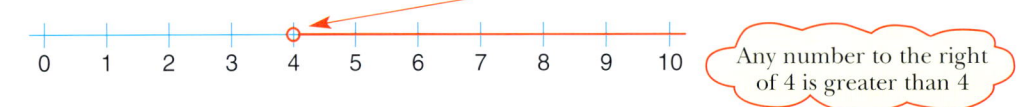

*The empty circle shows that x cannot be equal to 4*

*Any number to the right of 4 is greater than 4*

This is how you show $x \leq 2$.

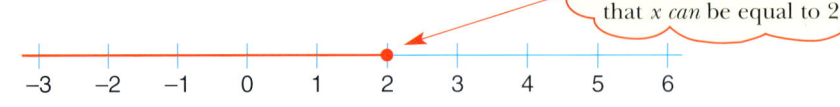

*The solid circle shows that x can be equal to 2*

*Any number to the left of 2 is less than 2*

**?** *Think of some possible values for x for each inequality and check that they come on the marked part of the number line.*

*Is –4 greater than 1 or less than 1?*

## Combining inequalities

'Young Singles' is a club for people in their twenties and thirties. The inequalities which describe the ages of its members are

$y \geq 20$ and $y < 40$.

*People who have had their 20th birthday are included*

As $y$ is *between* 20 and 40 you can combine these and write $20 \leq y < 40$.

You can show it on a number line like this.

*People who have had their 40th birthday are not included*

**?** *Write a similar inequality for the age of a student at your school or college.*

*Penny is 1.6 m tall (correct to 1 decimal place).*

*Write an inequality to show exactly what this means, and show it on a number line.*

To qualify for cheap rail tickets you must be under 16, or else 60 or over. Your age has to satisfy the inequality

$x < 16$ or $x \geq 60$.

You can show it on a number line like this.

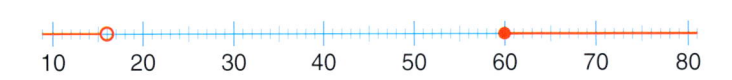

**?** *The number x has a square that is greater than 9. Where can x lie on the number line?*

88

# 7: Inequalities

**1** Show each of these inequalities on a number line.
   a) $x \geq 2$
   b) $x > -1$
   c) $x < 8$
   d) $x \leq 2$
   e) $x < 0$
   f) $x$ is positive
   g) $x$ is not more than 12
   h) $x$ is at least 6
   i) $x$ is not positive

**2** A second class stamp can be used to send a letter weighing no more than 60 g.
   a) Write this as an inequality for $w$, the weight of the letter in grams.
   b) Show your answer on a number line.

**3** Show these measurements on a number line.
   a) 3.8 m (correct to 1 d.p.)
   b) 2.0 kg (correct to 1 d.p.)
   c) 5 m (to the nearest metre).
   d) 3000 m (to the nearest 1000 m)
   e) 3000 m (to the nearest 100 m)
   f) 0.16 kg (correct to 2 d.p.)

**4** Combine the inequalities and show the results on a number line
   a) $x \geq 3$ and $x \leq 7$
   b) $x \leq 5$ and $x \geq -2$
   c) $x < 0$ and $x \geq -4$
   d) $y \geq 1.65$ and $y < 1.75$
   e) $a > -1$ and $a \leq 0$
   f) $b < -3$ and $b > -6$
   g) $x \geq -1$ and $x \geq 2$ (Be careful!)
   h) $c \leq 7$ and $c \leq 2$

**5** You join the queue at the supermarket checkout with this sign.

Write down an inequality for the number of items, $x$, in your basket.

Write down all the possible values of $x$ and mark these points on a number line.

EXPRESS CHECKOUT
FEWER THAN 10 ITEMS

**6** a) Copy and complete this table.

| $x$ | –5 | –4 | –3 | –2 | –1 | 0 | 1 | 2 | 3 | 4 | 5 |
|---|---|---|---|---|---|---|---|---|---|---|---|
| $x^2$ |  |  | 9 |  | 1 |  |  | 4 |  |  |  |

b) Show each of these on a number line, using your table to help you.
   (i) $x^2 \leq 16$
   (ii) $x^2 \geq 9$
   (iii) $x^2 < 25$
   (iv) $x^2 > 1$
   (v) $x^2 > 4$

---

Data has been collected to show the heights of members of your class. Inequality signs are used to show the class limits: $155 \leq h < 160$, $160 \leq h < 165$, etc. Draw a number line to show the range of possible values for an item in each class. Use a different colour for each one.

What do you notice?

# 7: Inequalities

## Solving inequalities

**?** For what values of x is  $3x - 5 > 16$?

You can solve this inequality in just the same way as you solve an equation.

*An equation*

$3x - 5 = 16$     $3x - 5 > 16$     *An inequality*

Add 5 to both sides    $3x = 21$     $3x > 21$

Divide both sides by 3    $x = 7$     $x > 7$     *This is the solution of the inequality. It means that x can have any value greater than 7.*

*The equation has just one solution*

*Check using a number just greater than 7*

Check:     When $x = 7$     When $x = 8$

$3x - 5 = 21 - 5 = 16$ ✓     $3x - 5 = 24 - 5 = 19$ ✓

*The inequality holds because 19 > 16*

**?** Check for yourself that when $x = 6$ the inequality is not true.

**?** Find the value of the expression $3x - 5$ when $x = 10$ and when $x = 7.5$.

What do you think will happen when $x = 2$?

You can do what you like to an equation, so long as you do the same to both sides. You can add or subtract any number, and you can multiply or divide by any number.

**?** Look at the inequality $30 > 20$. You know that it is valid.

Show that it is still valid if you add, subtract, multiply by or divide by 5.

Show that it is no longer valid when you multiply or divide by −5.

What happens to the inequality sign?

> **You solve inequalities just as you solve equations, except that if you multiply or divide by a negative number you must reverse the inequality sign.**

### Example

Solve $-2x > 8$

### Solution

$-2x > 8$

Divide by −2     $x < -4$

Check: when $x = -5$, $-2x = (-2) \times (-5) = 10$ ✓

*Another way to do this is to add 2x to both sides then subtract 8 from both sides. Try this to check you get the same answer.*

# 7: Inequalities

**Exercise**

**1** Solve these inequalities.
  a) $x + 1 < 7$
  b) $2x + 1 \leq 11$
  c) $x - 3 \geq 5$
  d) $3x - 2 \geq 10$
  e) $x - 4 < -1$
  f) $2x + 17 > 29$
  g) $5x - 2 \leq 16$
  h) $2x + 11 \leq 5$
  i) $3 + 4x > 11$
  j) $5 - x \geq 3$
  k) $7 - 2x \leq 1$
  l) $14 \leq 5 - 3x$

**2** Find all the possible values of $y$ when
  a) $y < 20$ and $y$ is a prime number
  b) $20 \leq y \leq 40$ and $y$ is a square number
  c) $3 < y < 12$ and $y$ is a factor of 12
  d) $12 > y > 3$ and $y$ is a multiple of 3.

**3** Last Saturday Grandad went to the races. He placed a bet of £$x$ on the first race and his horse came in first at 5 to 1. Grandad won £$5x$. After that he lost £200 of his winnings.

  a) Write an expression for the winnings he had left.
  b) This amount was still more than he had bet on the first horse.
     Write this as an inequality and solve it for $x$.

**4** Ben looks at the weight card his mother has kept from when he was a baby. He finds that he now weighs 52 kg more than he did when he was born. This is more than 14 times his birthweight.

Write this as an inequality for $w$, his birthweight, and solve it.

**5** Solve these inequalities.
  a) $2x + 5 \geq x + 16$
  b) $4x - 3 \geq x + 3$
  c) $8 < x + 1$
  d) $2(x + 4) < 20$
  e) $4 \leq x + 3 \leq 11$
  f) $20 \geq 2x > 6$
  g) $13 < 2x < 19$
  h) $0 \leq x - 2 \leq 18$
  i) $5 < 2x + 1 \leq 23$
  j) $3(x - 2) > 2x + 5$
  k) $10(x - 4) \geq 5(x + 8)$
  l) $2 - x > 3 - 2x$

---

Choose an inequality for $x$, such as $x < 5$ or $x \geq 1$.

Multiply both sides by the same number.

Add or subtract the same number on both sides.

You now have an inequality for a partner to solve.

Write it on another piece of paper so your working is not visible.

See if your partner can solve the inequality and get back to your starting point.

# 7: Inequalities

## Inequalities and graphs

Look at this chart showing people's heights and weights.

If you plot your height and weight on it you can see whether you are overweight, underweight or about right for your height.

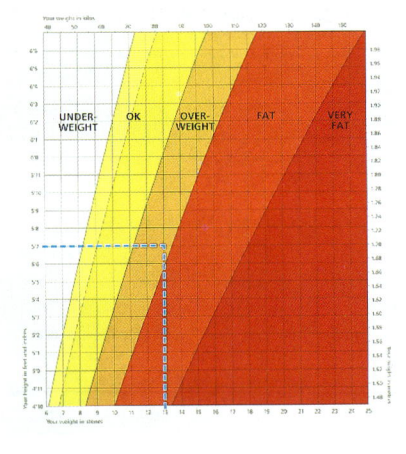

In this situation you are content if the point you plot is in the central **region** of the graph. You do not need it to be on a particular line.

You can often describe regions of a graph using inequalities. In this chapter you will learn how to use inequalities to describe some regions, and how to show on a graph the region described by one or more inequalities.

Look at this graph. It shows the region for which

$0 \leq y \leq 3$ and $0 \leq x \leq 5$.

All points in the shaded region have $x$ and $y$ values that satisfy these inequalities.

**?** *Check that this is true for point A.*

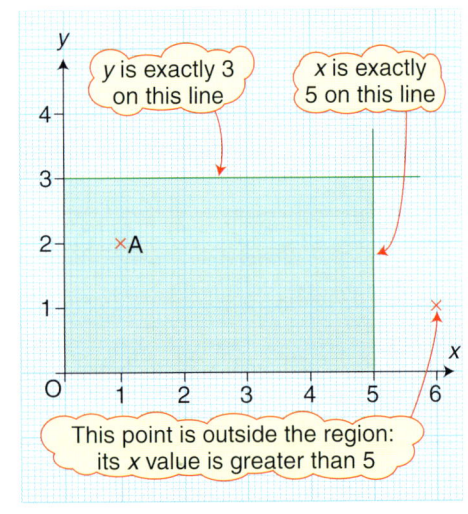

$y$ is exactly 3 on this line

$x$ is exactly 5 on this line

This point is outside the region: its $x$ value is greater than 5

This graph shows the region for which

$3 < x \leq 5$ and $y \geq 1$.

**?** *Is point P inside the region? Explain your answer.*

**?** *Sketch a graph showing the region for which $3 < x \leq 5$ and $y \leq 0$.*

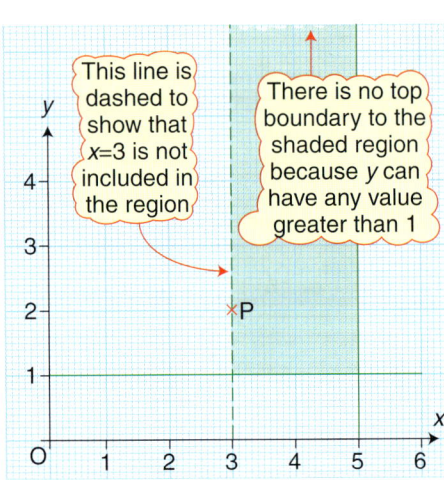

This line is dashed to show that $x=3$ is not included in the region

There is no top boundary to the shaded region because $y$ can have any value greater than 1

92

# 7: Inequalities

**1** Write down the inequalities represented by the shaded regions.

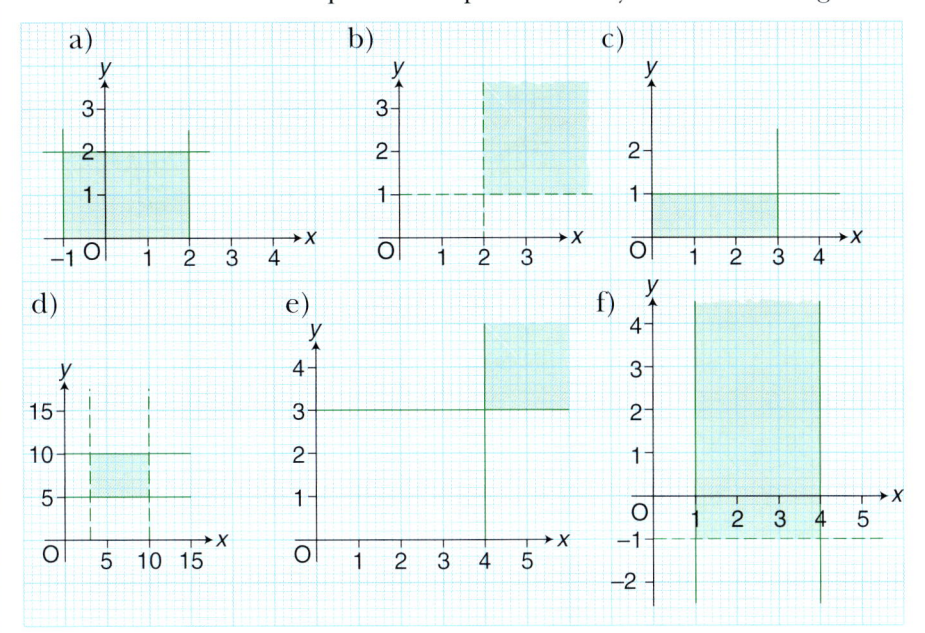

a) b) c)

d) e) f)

**2** Sketch graphs and shade the regions that represent these inequalities.

a) $0 \leq x \leq 4$
   $0 \leq y \leq 8$

b) $-1 \leq x \leq 3$
   $2 < y < 4$

c) $x \geq 2$
   $y > 1$

d) $x < 2$
   $0 \leq y < 4$

e) $30 \leq x < 50$
   $20 \leq y < 40$

f) $x \geq 1$
   $y \leq 5$

**3** In a hot air balloon race the winners were in the air for 2 hours and went a distance of 100 miles. They went furthest and were in the air for the longest time.

The shortest distance travelled by any of the balloons was 5 miles and the shortest time in the air was 20 minutes.

a) Write down suitable inequalities for the time taken and the distance travelled by the competitors in the race.

b) Illustrate these inequalities on a graph and shade the region in which all the balloons' times and distances could be plotted.

# 7: Inequalities

## Regions bounded by sloping lines

So far you have met regions of a graph **bounded** by horizontal and vertical lines, but regions can be bounded by sloping lines.

Look at this graph of

$y = x + 1$

The red points are above the line and the blue ones are below it.

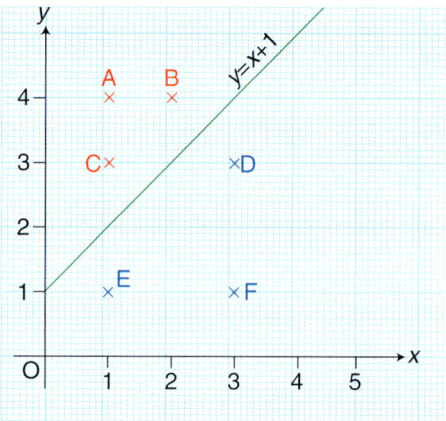

**?** *For each point, write down the co-ordinates (x, y), then write an inequality of the form $y > x + 1$ or $y < x + 1$. What do you notice?*

### Example

Draw the graph of $x + y = 10$.

Shade the region in which

$x + y \leq 10$, $y \geq 0$ and $x \geq 0$.

### Solution

Make a table of values for $x + y = 10$.

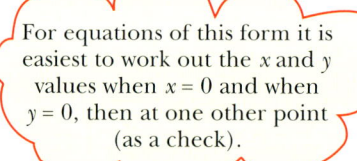

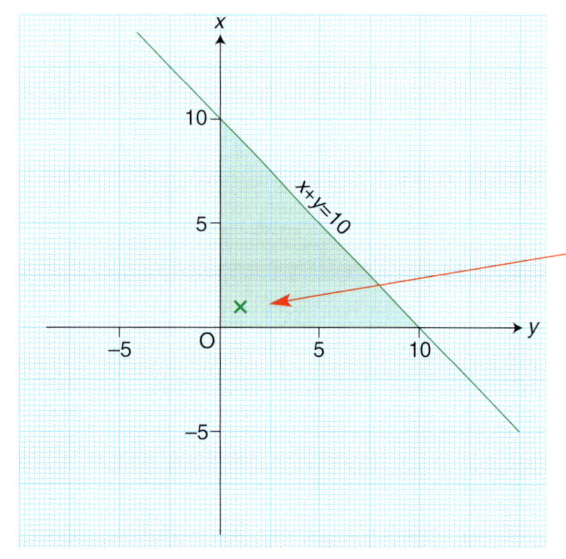

| x | 0  | 10 | 5 |
|---|----|----|---|
| y | 10 | 0  | 5 |

*For equations of this form it is easiest to work out the x and y values when $x = 0$ and when $y = 0$, then at one other point (as a check).*

*Test point at (1,1): $x + y = 2$. On this side of the line, $x + y < 10$*

**?** Which region would you shade for $x + y \leq 10$, $x \geq 0$ and $y \leq 0$?

Which region would you shade for $x + y \leq 10$, $x \leq 0$ and $y \leq 0$?

# 7: Inequalities

**1** a) Draw a graph of the line $x + y = 6$ for values of $x$ and $y$ from 0 to 7.

b) Mark these points on your graph:

A(3, 6), B(5, 5), C(6, 2), D(1, 3), E(2, 2) and F(4, 1).

c) Find the value of $x + y$ at each of the points A to F.

d) Shade the region of your graph where $x + y \leq 6$.

**2** a) Draw a graph of the line $y = 2x + 2$ for values of $x$ from 0 to 5 and of $y$ from 0 to 12.

b) Mark these points on your graph:

P(0, 4), Q(2, 8), R(3, 9), S(2, 1), T(5, 1).

c) Find the value of $2x + 2$ at each of the points P to T.

d) Shade the region of your graph where $y \leq 2x + 2$.

**3** a) Draw a graph of the line $y = x$ for values of $x$ and $y$ from 0 to 5. (Use a dashed line.)

b) Draw on your graph the line $y = 4$ (again dashed, not solid).

c) By using test points, find the region of your graph where $y > x$ and $y < 4$.

d) Why was it important to use dashed lines?

**4** Sanjay is making a table. He has a carved strip 500 cm long that he wants to put round the edge of the table. The table is to be rectangular, of width $x$ and length $y$ cms.

a) Write down an expression for the perimeter of the table in terms of $x$ and $y$.

b) Explain why $2x + 2y \leq 500$ and write this inequality in its simplest form.

c) Sanjay wants the table to be more than 80 cm wide. Write this as an inequality.

d) Draw a graph and shade the feasible region for the dimensions of the table.

**5** For each of these, draw a graph and shade the region represented by the inequalities.

a) $0 < x < 6$, $3 < y < 6$

b) $0 \leq x \leq 4$, $2 \leq y \leq 6$, $x \geq y$

c) $x + y \geq 4$, $0 \leq x \leq 8$, $0 \leq y \leq 10$

d) $y < x + 3$, $y \geq 0$, $1 < x < 5$

# 7: Inequalities

## Solution sets

Jackie has a £12 gift token to spend. She wants to use it to buy more mugs and plates like these for her bedsit.

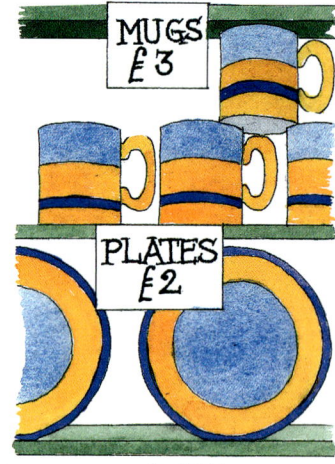

**?** *Find a combination of mugs and plates that costs exactly £12.*

*Find a combination that costs less than £12.*

Jackie can buy any combination that costs up to £12.

You can write this as an inequality:

$3x + 2y \leq 12$

- *x* is the number of mugs. $3x$ is their cost in pounds.
- *y* is the number of plates. $2y$ is their cost in pounds.

**?** *The numbers x and y must be integers in this problem. Why?*

You can show the solutions to this inequality on a graph. You start by drawing the line

$3x + 2y = 12$

Then use a test point to decide on which side of the line $3x + 2y \leq 12$.

The blue crosses show all the combinations that Jackie can buy. These points make up the **solution set**.

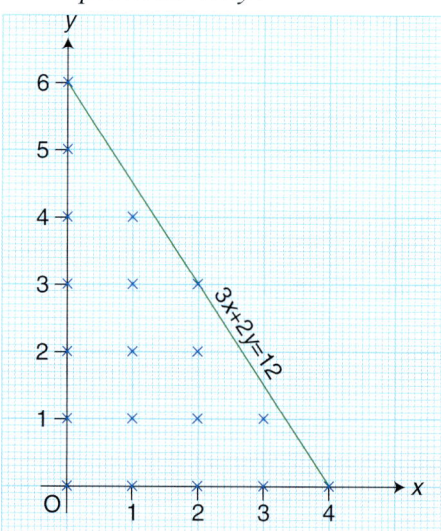

**?** *Why are none of the points to the right of the line?*

*Why is it not correct to shade the whole region to the left of the line in this case?*

*Points on the x and y axes are included in the solution. Why?*

*Points for which x or y is negative are not included. Why not?*

*Three of the points in the solution set lie on the line $3x + 2y = 12$. What is special about these points?*

# 7: Inequalities

**1**  a) Draw a graph of $x + 3y = 9$ for $0 \leq x \leq 10$ and $0 \leq y \leq 4$.

b) Find the points (1, 2) and (2, 4) and work out the value of $x + 3y$ at each of these points.

c) Shade the region where $x + 3y \leq 9$.

d) Mark the set of points where $x$ and $y$ are both positive integers and where $x + 3y \leq 9$.

**2**  Kim is a gymnast. She has to train for at least 10 hours a week.

There are two training sessions on a Saturday, each 4 hours long, and the weekday sessions are 2 hours each evening. There is no training on Sundays.

a) Using $w$ for the number of weekday sessions and $s$ for the number of Saturday sessions that Kim attends in a week, write an inequality for $w$ and $s$.

b) Clearly Kim cannot attend more than 5 weekday sessions in a week. Write this as an inequality for $w$.

c) Write a similar inequality for $s$.

d) Show all the possible combinations of $s$ and $w$ on a graph.

**3**  Dean works for $x$ hours during the week and $y$ hours at the weekend. He cannot work for more than 21 hours per week because he is at college.

a) Write down an inequality to show this.

b) Dean earns £4 per hour during the week and £7 per hour at weekends. He aims to earn at least enough to pay his rent of £56 a week. The shifts he works are always whole numbers of hours.

Write down a second inequality to show this.

c) Draw a suitable graph and mark on it the points in the solution set.

d) Dean tries to do as much weekend work as possible because it pays better. He is told that the number of weekend hours must not be more than the number of weekday hours.

Add the line $y = x$ to your graph.

e) Using algebra, find the co-ordinates of the point where $y = x$ crosses $x + y = 21$.

f) Find the maximum amount Dean can earn in a week.

# 7: Inequalities

## Finishing off

**Now that you have finished this chapter you should**

★ understand the symbols <, >, ≤, ≥

★ be able to represent an inequality on a number line

★ be able to solve an inequality using algebra

★ be able to represent inequalities for $x$ and $y$ by shading areas or marking sets of points on a graph.

Use the following questions to check that you understand everything.

## Mixed exercise

**1** Write each of these inequalities in words and show it on a number line.

a) $x > 2$   b) $x \leq 1$   c) $x \geq 4$

d) $x < -3$   e) $-2 < x < 3$   f) $1 \leq x \leq 5$

**2** Write the age ranges in these lonely hearts' advertisements as inequalities. Show each one on a number line.

- **M 5′ 9″** dark, lively personality, likes jazz and dancing, Seeks F 20–35, for friendship and outings.
- **F** professional, young 36 seeks M 30s for companionship/romance.

**3** Madeleine has inherited some money from her aunt.

She puts it in one of these accounts.

It is earning 5.70% annual interest.

Write an inequality for the sum, £$m$, that Madeleine inherited.

**RAINBOW Building Society — SAVINGS RATES**

| Balance | Annual Interest |
|---|---|
| £1 – £4,999 | 5·35% |
| £5,000 – £9,999 | 5·70% |
| £10,000 – £24,999 | 6·20% |
| £25,000 – £49,999 | 6·50% |
| £50,000 – £99,999 | 6·70% |
| £100,000 + | 6·80% |

**4** Solve these inequalities.

a) $x + 3 \leq 8$

b) $2x - 5 \geq 21$

c) $2x + 1 > x + 9$

d) $5(x + 11) < 80$

e) $2(2x - 3) \leq 3x + 2$

f) $-3 < x + 1$

g) $4(x + 2) - 3(x + 1) > 12$

h) $-5x \geq 25$

i) $9 < x - 2 < 12$

# 7: Inequalities

**Mixed exercise**

**5** Write down the inequalities represented by the shaded areas.

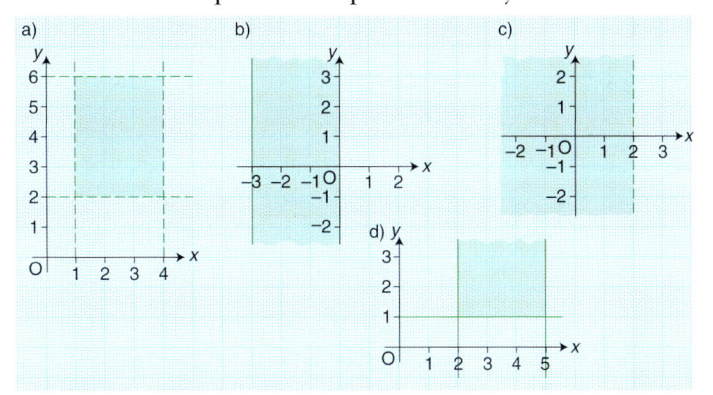

a) b) c) d)

**6** Draw suitable graphs and shade the regions that represent the following inequalities.

a) $1 \leq x \leq 6$
$-1 \leq y \leq 5$
$x \leq y$

b) $x + y \leq 10$
$x \geq 0$
$y \geq 0$

c) $y > x + 2$
$0 \leq x \leq 5$
$y < 9$

d) $y \geq 0$
$x \geq 1$
$y \geq 2x$

e) $-2 \leq x \leq 2$
$-3 \leq y \leq 2$
$x + y \leq 0$

f) $-3 \leq x \leq 3$
$0 < y < 5$
$y > 2x + 1$

**7** In a pub quiz, a team scores 10 points for a correct answer to a starter question. If they get the starter question right, they are asked a bonus question worth a further 5 points.

a) How many points does a team score when it gets 30 starter questions and 19 bonus questions right?

b) Write down an expression for the score when the team gets $x$ starter questions and $y$ bonus questions right.

c) Last season James's team's highest score was 430 points.

Write down an inequality that is true for each of his team's scores last season.

d) Explain why $x \geq 0$, $y \geq 0$ and $x \geq y$.

e) Draw a graph to illustrate the inequalities in c) and d) and show where you could mark the points in the solution set.

**8** a) Draw a graph of the line $3x + 4y = 60$.

b) On the same axes, draw the line $x + y = 17$.

c) Shade the region in which

$x \geq 0$, $y \geq 0$, $3x + 4y \leq 60$ and $x + y \geq 17$.

d) Write down the co-ordinates of the vertices of your shaded triangle.

99

# Eight

# Indices and standard form

**Before you start this chapter you should be able to**

★ use simple index notation, such as $4^3$, $2^5$, ...

★ find squares, cubes and higher powers

★ find square roots and cube roots

★ multiply and divide numbers by powers of 10 without using a calculator

★ work out the value of a large number given in standard form on a calculator display.

## Reminder

Multiplying by 10 moves the decimal point one place to the right.

$4.5 \times 10 = 45$

Dividing by 10 moves the decimal point one place to the left.

$243 \div 10 = 24.3$

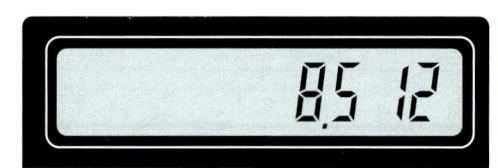

> A display like this (or similar to it) means $8.5 \times 10^{12}$ which is 8 500 000 000 000

## Revision exercise

**1** Work out the value of

a) $2^3$  b) $6^2$  c) $3^4$  d) $2^6$  e) $3^5$  f) $4^4$  g) $11^2$  h) $1.5^3$

**2** Work out the value of

a) $9^2$  b) $\sqrt{81}$  c) $5^3$  d) $\sqrt[3]{125}$

e) $20^3$  f) $\sqrt{49}$  g) $\sqrt[3]{1000}$  h) $400^2$

**3** Jody packs sugar cubes in boxes like this.

a) How many are there on the top layer?

b) How many are there in the box?

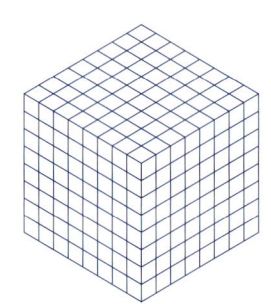

# 8: Indices and standard form

**Revision exercise**

**4** Steven is designing a garden. He has 200 square slabs. He uses all these slabs to make two identical square patios.

   a) How many slabs are there along an edge of one of these patios?

   Steven decides that he would rather have one large patio.

   b) What size is the largest square patio that he can make?

   c) How many slabs are left over?

**5** Work out

   a) $52 \times 10$
   b) $83 \div 10$
   c) $6.9 \div 10$
   d) $23 \times 100$
   e) $47 \div 100$
   f) $6.4 \times 10$
   g) $0.7 \times 10$
   h) $528 \div 100$
   i) $93.45 \times 100$
   j) $0.8 \times 10$
   k) $9.12 \div 10$
   l) $573 \times 1000$
   m) $7.5 \div 100$
   n) $39\,371 \div 100$
   o) $14.065 \times 10$
   p) $105\,600 \div 10$

**6** Work out the value of the numbers displayed here.

   a) 2.8 11

   b) 5.12 13

## Investigation

Copy and complete this table.

| | | | |
|---|---|---|---|
| $2^1 = 2$ | $4^1 =$ | $6^1 =$ | $8^1 = 8$ |
| $2^2 =$ | $4^2 =$ | $6^2 = 36$ | $8^2 =$ |
| $2^3 =$ | $4^3 = 64$ | $6^3 =$ | $8^3 =$ |
| $2^4 = 16$ | $4^4 =$ | $6^4 =$ | $8^4 = 4096$ |
| $2^5 =$ | $4^5 =$ | $6^5 = 7776$ | $8^5 =$ |

Look at the pattern of last digits of the numbers in each column.

(i) What do you think the last digit is for each number to the power 6?

(ii) Write down what you think the last digit is in each of these numbers.

   a) $4^9$   b) $6^8$   c) $2^{11}$   d) $8^7$   e) $2^9$   f) $8^{10}$

Check that you can use your calculator to find cube roots.

# 8: Indices and standard form

# Rules of indices

Look at this table.

| | |
|---|---|
| $16 = 2 \times 2 \times 2 \times 2$ | $2^4$ |
| $8 = 2 \times 2 \times 2$ | $2^3$ |
| $4 = 2 \times 2$ | $2^2$ |
| $2 = 2$ | $2^1$ |
| $1 = 1$ | $2^0$ |
| $\frac{1}{2} = \frac{1}{2}$ | $2^{-1}$ |
| $\frac{1}{4} = \frac{1}{2 \times 2} = \frac{1}{2^2}$ | $2^{-2}$ |
| $\frac{1}{8} = \frac{1}{2 \times 2 \times 2} = \frac{1}{2^3}$ | $2^{-3}$ |

(÷ 2 arrows on both sides)

You can see the meaning of 2 to the power zero and to a negative power. Notice that

$$2^0 = 1 \qquad \text{and} \qquad 2^{-3} = \frac{1}{2^3}$$

**?** What are the values of $10^0$ and $10^{-2}$?

When you write 16 as $2^4$ then $2^4$ is called **index form**.

When you multiply two numbers given in index form you add the powers.

$$2^4 \times 2^3 = 2^{4+3} = 2^7$$

$2^4 \times 2^3 = (2 \times 2 \times 2 \times 2) \times (2 \times 2 \times 2) = 2^7$

**?** What is $2^2 \times 2^3 \times 2^4$?

When you divide one number by another you subtract the powers.

$$2^5 \div 2^3 = 2^{5-3} = 2^2$$

$\frac{2^5}{2^3} = \frac{2 \times 2 \times 2 \times 2 \times 2}{2 \times 2 \times 2} = 2 \times 2 = 2^2$

**?** What is $2^4 \div 2$? (Hint: 2 is the same as $2^1$).

**?** What is $(2^3)^4$ in index form?

You can write the rules on this page as general laws.

$$a^0 = 1 \qquad\qquad a^{-n} = \frac{1}{a^n}$$

$$a^m \times a^n = a^{m+n} \qquad\qquad a^m \div a^n = a^{m-n}$$

# 8: Indices and standard form

**Exercise**

**1** Write each value as a fraction. For example $2^{-2} = \frac{1}{4}$

a) $4^{-2}$  b) $10^{-3}$  c) $5^{-2}$  d) $8^{-1}$  e) $3^{-3}$  f) $6^{-2}$

g) $10^{-2}$  h) $3^{-4}$  i) $4^{-1}$  j) $2^{-4}$  k) $6^{-3}$  l) $10^{-4}$

**2** Work out the value of

a) $7^2$  b) $3^{-2}$  c) $10^{-1}$  d) $4^3$  e) $9^{-2}$  f) $6^0$

g) $2^5$  h) $5^{-3}$  i) $10^6$  j) $9^1$  k) $5^{-1}$  l) $7^0$

m) $6^3$  n) $8^{-2}$  o) $4^1$  p) $5^4$

**3** Work these out giving your answer in index form.

For example $4^3 \times 4^2 = 4^5$

a) $5^2 \times 5^4$  b) $2^6 \times 2^3 \times 2^2$  c) $6 \times 6^3$  d) $10^6 \div 10^3$

e) $2^8 \div 2$  f) $(10^2)^3$  g) $(3^4)^2$  h) $4^5 \times 4^{-3}$

i) $10^3 \div 10^{-1}$  j) $(5^{-1})^2$  k) $3^4 \times 3^{-2} \times 3$  l) $2^{-1} \div 2^{-2}$

m) $\dfrac{4^9}{4^3 \times 4^4}$  n) $\dfrac{3^4 \times 3^2}{3^8}$  o) $\dfrac{(2^3)^2 \times 2}{2^7}$  p) $\dfrac{10^4 \times 10^6 \times 10}{10^5 \times 10^3}$

## Investigation

**1** Use your calculator to work out

a) $7 \times 10$

b) $7 \times 10 \times 10$

c) $7 \times 10 \times 10 \times 10$

. . . and so on until you get an answer in standard form.

How many digits can your calculator display?

**2** Use your calculator to work out

a) $7 \div 10$

b) $7 \div 10 \div 10$

c) $7 \div 10 \div 10 \div 10$

. . . until the form of answer changes.

For which calculation does the form of your answer change?

What does the display show?

What do you think this display means?

### 8: Indices and standard form

# Calculations using standard form

This diagram shows the distances of the planets from the Sun.

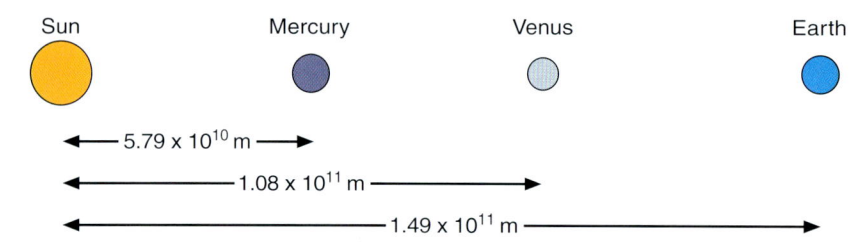

Sun       Mercury       Venus       Earth

←— $5.79 \times 10^{10}$ m —→

←——— $1.08 \times 10^{11}$ m ———→

←————— $1.49 \times 10^{11}$ m —————→

How far is Earth from Mercury when they are at their closest?

It is $(1.49 \times 10^{11} - 5.79 \times 10^{10})$ metres.

You work it out on your calculator like this.

*$1.49 \times 10^{11}$ is keyed in like this*

[ 1 ] [ . ] [ 4 ] [ 9 ] [EXP] [ 1 ] [ 1 ] [ − ] [ 5 ] [ . ] [ 7 ] [ 9 ] [EXP] [ 1 ] [ 0 ] [ = ]

*$5.79 \times 10^{10}$ is keyed in like this*

**Earth is $9.11 \times 10^{10}$ m from Mercury when they are at their closest.**

The volume of the Earth is $1.08 \times 10^{21}$ m$^3$.

Each cubic metre has, on average, a mass of $5.52 \times 10^3$ kg.

*Earth has a **density** of $5.52 \times 10^3$ kg/m$^3$*

What is the mass of the Earth?

> **mass = volume × density**

$$= (1.08 \times 10^{21}) \times (5.52 \times 10^3)$$

**?** Work this out on your calculator. Check you get $5.96 \times 10^{24}$.

**The mass of the Earth is $5.96 \times 10^{24}$ kg.**

**?** $5.96 \times 10^{24}$ is in standard form. What is its value?

The Earth is about $1.5 \times 10^{11}$ metres from the Sun.

Light travels at a speed of $3.0 \times 10^8$ metres per second.

How long does it take light to travel from the Sun to the Earth?

Kylie works it out like this:

$$\frac{\text{Time in}}{\text{seconds}} = \frac{\text{distance}}{\text{speed}} = \frac{1.5 \times 10^{11}}{3 \times 10^8} = \frac{1.5 \times 10^3}{3} = 500$$

**?** Check this calculation on your calculator.

What is 500 seconds in minutes and seconds?

# 8: Indices and standard form

**1** Work out these calculations, and give your answers in standard form.

a) $(4 \times 10^3) + (8 \times 10^4)$
b) $(7 \times 10^6) - (2 \times 10^5)$
c) $(3 \times 10^5) \times (5 \times 10^6)$
d) $(8 \times 10^{11}) \div (5 \times 10^4)$
e) $(2.5 \times 10^{-4}) \times (6 \times 10^{12})$
f) $(7.4 \times 10^{-2}) - (3.8 \times 10^{-3})$
g) $(4.5 \times 10^{-7}) \times (3.2 \times 10^{-11})$
h) $(4 \times 10^7)^2$

**2** Complete this table by multiplying volume by density to get mass.

| Planet | Volume (m$^3$) | Density (kg/m$^3$) | Mass (kg) |
|---|---|---|---|
| Mercury | $1.45 \times 10^{19}$ | $5.42 \times 10^3$ | |
| Venus | $9.29 \times 10^{20}$ | $5.25 \times 10^3$ | |
| Mars | $5.21 \times 10^{19}$ | $3.94 \times 10^3$ | |
| Jupiter | $1.56 \times 10^{24}$ | $1.31 \times 10^3$ | |

**3** The kinetic energy, $E$, of an electron is $\frac{1}{2}mv^2$.

When $m = 9 \times 10^{-31}$ kg and $v = 2 \times 10^7$ metres per second, what is $E$?

**4** This table shows the population and area of different continents.

| Continent | Population | Area (km$^2$) |
|---|---|---|
| Europe (incl Russia) | $4.95 \times 10^8$ | $4.94 \times 10^6$ |
| Africa | $7.43 \times 10^8$ | $3.03 \times 10^7$ |
| Oceania | $8.51 \times 10^6$ | $2.5 \times 10^7$ |

a) Population density = population ÷ area.

Which of these continents has the highest population density?

b) What is the population of Oceania to the nearest million?

c) What is the area of Africa to the nearest million km$^2$?

**5** The radius, $r$, of the moon is $1.7 \times 10^6$ metres.

Its volume, $V$, is $\frac{4}{3}\pi r^3$.

a) What is the volume of the moon?

b) The mass of the moon is $7.4 \times 10^{22}$ kg.

What is the density of the moon in kg/m$^3$?

---

Find out the radius of each planet in metres.
Set up a spreadsheet to work out the volume of each planet in m$^3$.

## 8: Indices and standard form

# *Finishing off*

**Now that you have finished this chapter you should be able to**

★ work out powers and roots of numbers
★ work out the value of numbers in standard form
★ write numbers in standard form
★ do calculations with numbers in standard form.

Use the questions in the next exercise to check that you understand everything.

**Mixed exercise**

**1** Work out

a) $8^2$  b) $10^3$  c) $6^{-1}$  d) $4^0$
e) $2^7$  f) $\sqrt{36}$  g) $12^2$  h) $\sqrt[3]{8}$
i) $3^{-2}$  j) $6^4$  k) $5^0$  l) $4^{-3}$
m) $\sqrt{225}$  n) $7^{-2}$  o) $\sqrt[3]{64}$  p) $2^{10}$

**2** Work these out, giving your answer in index form.
For example $3^2 \times 3^5 = 3^7$

a) $2^5 \times 2^3$  b) $7^4 \div 7$  c) $(5^3)^2$  d) $4^2 \times 4^6$
e) $6^5 \div 6^2$  f) $(4^2)^2$  g) $3^5 \times 3 \times 3^{-2}$  h) $\sqrt{3^2}$

**3** The numbers in this question are in standard form. Write them out in full.

a) The radius of the Earth is **6.4 × 10⁶** metres
b) A capillary tube has radius **2 × 10⁻⁴** metres
c) The wavelength of mercury green light is **5.4 × 10⁻⁷** metres
d) The density of mercury is **1.36 × 10⁴** kg/m³

**4** Write these numbers in standard form.

a) A train has a mass of **200 000** kg.
b) The thickness of a piece of cardboard is **0.0015** metres.
c) Thorium-230 has a half life of **83 000** years.
d) The linear expansivity of aluminium is **0.000 026** per degree Kelvin.

# 8: Indices and standard form

**Mixed exercise**

**5** Work out the value of these calculations, and give your answer in standard form.

a) $(5 \times 10^4) + (8 \times 10^5)$
b) $(3.1 \times 10^{-2}) - (7 \times 10^{-3})$
c) $(4 \times 10^8) \times (9.7 \times 10^{13})$
d) $(3.6 \times 10^{12}) \div (9 \times 10^4)$
e) $(3.2 \times 10^{14}) \times (7.5 \times 10^{-9})$
f) $(4.9 \times 10^{11}) \div (2.8 \times 10^{-5})$
g) $(4.5 \times 10^7)^2$
h) $\sqrt{1.6 \times 10^{13}}$

**6** The diameter of the Earth is $1.28 \times 10^7$ metres.

Work out the distance around the equator in metres.

**7**
a) Calculate the number of seconds in a year.

b) Light travels at $3 \times 10^8$ metres per second.

How many metres does light travel in a year?

*A light year is the distance travelled by light in a year*

c) How many kilometres is this?

d) Sirius A is a star 4.2 light years away from Earth. How many kilometres is this?

## Investigation

When you listen to a radio station you must tune in to the appropriate frequency. The frequency is often written in kilohertz (kHz). It is sometimes written in Hertz (Hz) and then the wavelength, $\lambda$, is given by $\lambda = v/f$ where $f$ is the frequency.

*v stands for the speed of the radio waves*

Copy and complete this table. (Note. $f$ is required in kHz in the table but must be in Hz to use in the formula, and $v = 3 \times 10^8$ ms$^{-1}$.)

| Station | $f$(kHz) | $f$(Hz) | $\lambda$(m) |
|---|---|---|---|
| Radio 4 | 198 | | |
| Radio 4 | | | 417 |
| Talk Radio | 1089 | | |
| Radio 5 Live | | | 433 |
| Radio 5 Live | 909 | | |
| Virgin | | | 247 |
| World Service | 648 | | |

Find out what these prefixes mean:
a) mega-   b) micro-   c) giga-   c) nano-

Try to find examples of when they are used in real life.

# Nine

# Circles and tangents

**Before you start this chapter you need to know that**

- ★ angles on a straight line add up to 180° and those round a point add up to 360°
- ★ the angle sum of a triangle is 180° and that of a quadrilateral is 360°
- ★ the base angles of an isosceles triangle are equal.

## Shapes in a circle

ABCD is a quadrilateral. It is called a **cyclic quadrilateral** because all 4 points are on the circumference of a circle.

**?** *What is the angle sum of a cyclic quadrilateral? What is the size of the angle at D?*

*The angles at A and C add up to 180°*

*So do the angles at B and D*

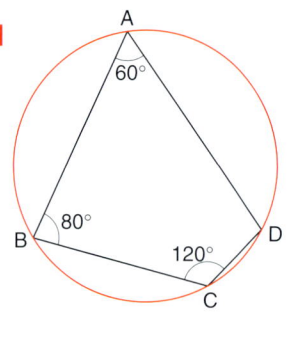

Draw a circle and a cyclic quadrilateral of your own.
Measure all the angles and add the pairs of opposite angles.

**?** *What do you notice? Does everyone get the same result?*

**?** *Complete this statement: 'Opposite angles of a cyclic quadrilateral … .'*

**?** *What is the value of a? What is the value of b?*

Here is a triangle drawn in a circle.
AB is the diameter.

Measure the angle at C.

Draw your own circle and add a diameter.

Complete a triangle by joining the ends of the diameter to any point on the circumference.

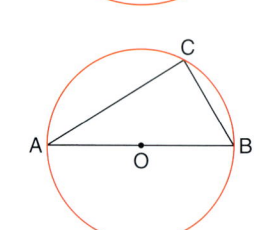

Measure the angle at the circumference.

**?** *What do you notice? Does everyone get the same result?*

**?** *Complete this statement: 'The angle in a semi-circle is … .'*

# 9: Circles and tangents

**1** The diagram shows a circle, with centre O. AB is a diameter of the circle and C is a point on the circumference. OA, OB and OC are all the radii of the circle, so the triangles OAC and OBC are both isosceles.

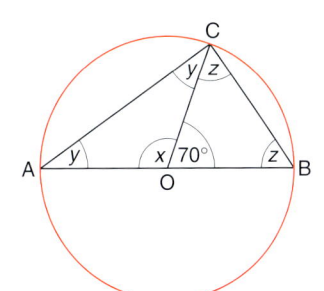

  a) (i) Find the angles marked with letters.

  (ii) Find the angle ACB by adding together angles y and z.

  b) Choose a different angle instead of 70° for angle COB, and repeat part a) of this question. Do this several times.

  c) What have you found out about the angle ACB?

**2** Check that your answer to question 1 c) is correct before you start this question.

You will need to use the rule that you have found.

In each of these diagrams, PQ is a diameter of the circle.

Find the angles *a* and *b*.

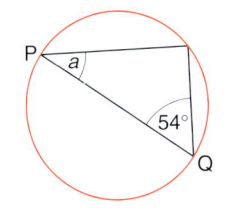

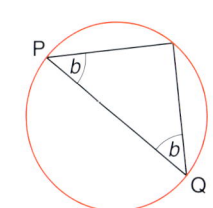

**3** Find the angles marked with letters in these diagrams. Give reasons for your answers.

a)

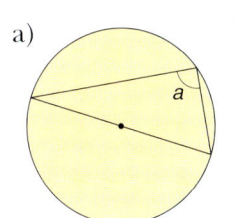

b) 66°

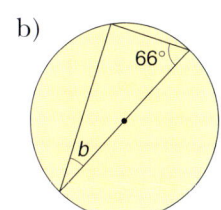

c) 78°

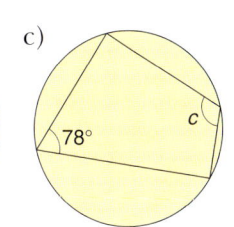

d) 141°

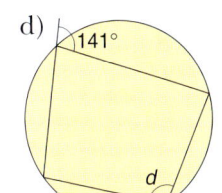

e) 34°

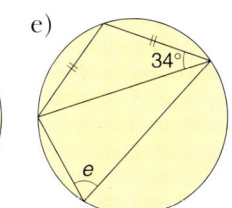

f) 108°, 82°

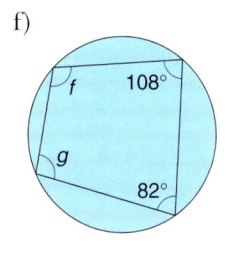

g) 110°

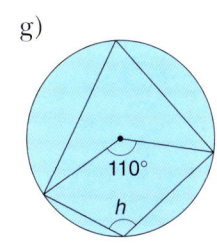

h) 28°

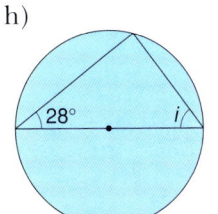

i) 94°

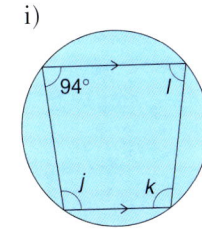

j) 88°

109

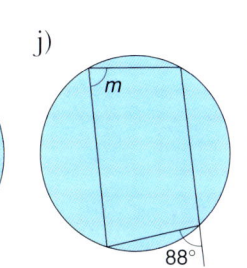

# 9: Circles and tangents

## Angles in a circle

These two angles are standing on the same arc AB.

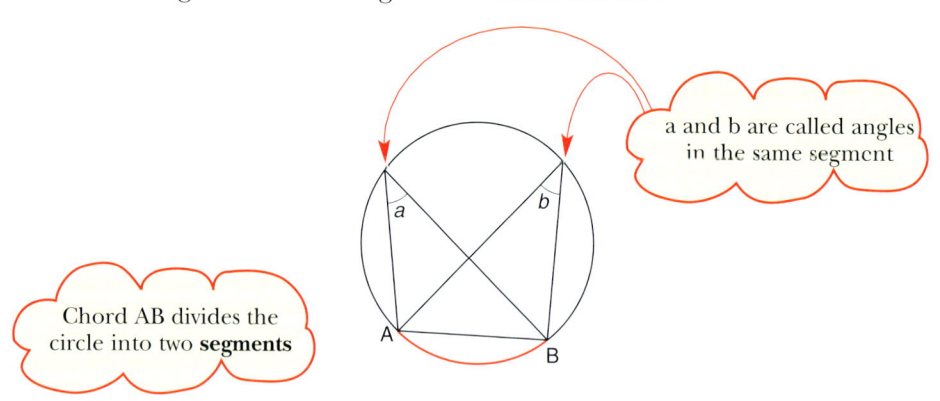

*a and b are called angles in the same segment*

*Chord AB divides the circle into two **segments***

**?** *Measure angles a and b. What do you notice?*

**?** *Are there any other equal angles in the diagram?*

**?** *Complete this statement: 'Angles in the same segment … .'*

**?** *Which angles are equal in this diagram?*

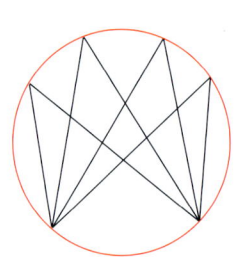

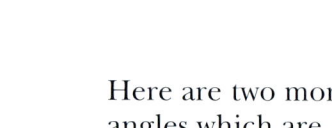

Here are two more angles which are standing on the same arc AB.

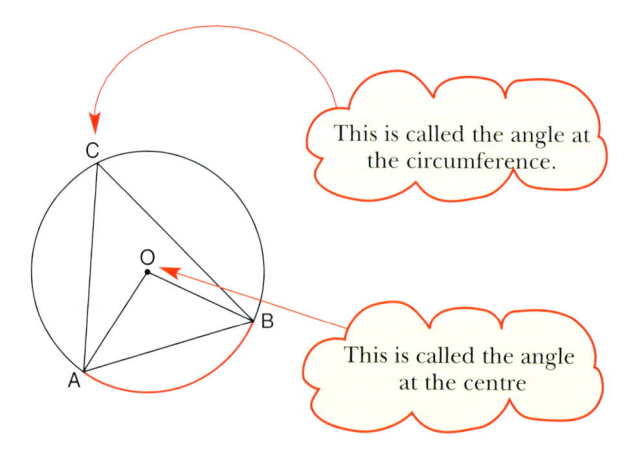

*This is called the angle at the circumference.*

*This is called the angle at the centre*

Measure the angle at the centre and the angle at the circumference.  *One is twice the other*

Draw a similar diagram of your own.
Measure the angle at the centre and the angle at the circumference.

**?** *What do you notice? Does everyone get the same result?*

**?** *Complete this statement: 'The angle at the centre of a circle is … .'*

# 9: Circles and tangents

**1** a) Explain why triangle AOC is isosceles.
b) Find *a*.
c) Find the third angle of triangle AOC.
d) Find *x*.
e) Use a similar method to find *y*.
f) What size is (i) AOB (ii) ACB?
g) What is the connection between these two angles?

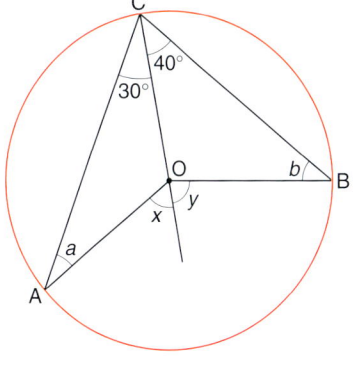

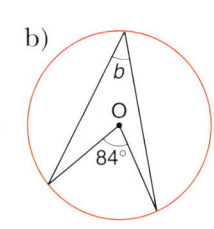

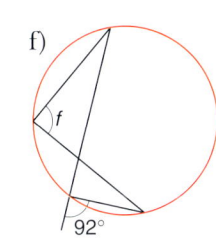

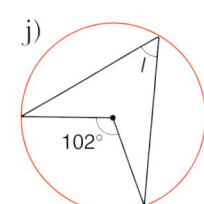

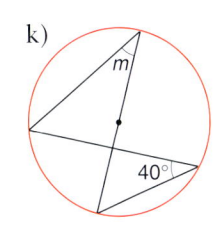

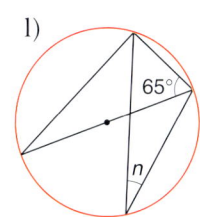

**2** Find the angles marked with letters in these diagrams. Give reasons for your answers.

a) 36° , a
b) b , 84°
c) 128° , c 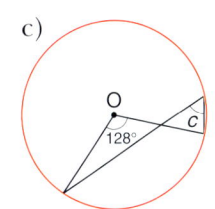
d) d , 28°
e) 31° , 114° , e 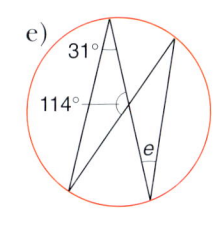
f) f , 92°
g) 140° , g 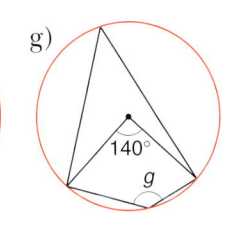
h) h , 35°
i) i , 25° , j , k
j) l , 102°
k) m , 40°
l) 65° , n
m) 28° , o , p , q
n) 48° , r

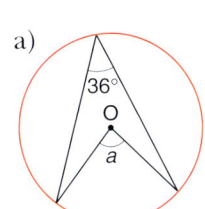

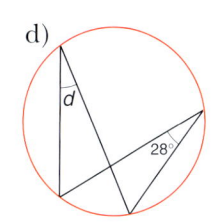

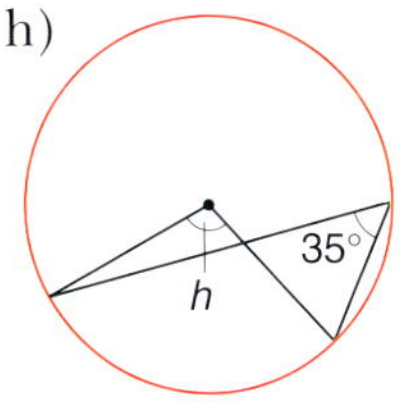

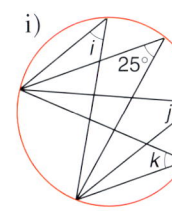

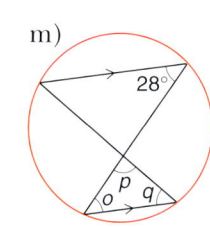

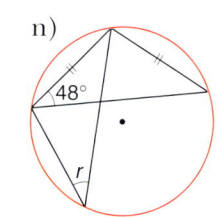

## 9: Circles and tangents

# Tangents

A **tangent** is a line which just touches a circle.

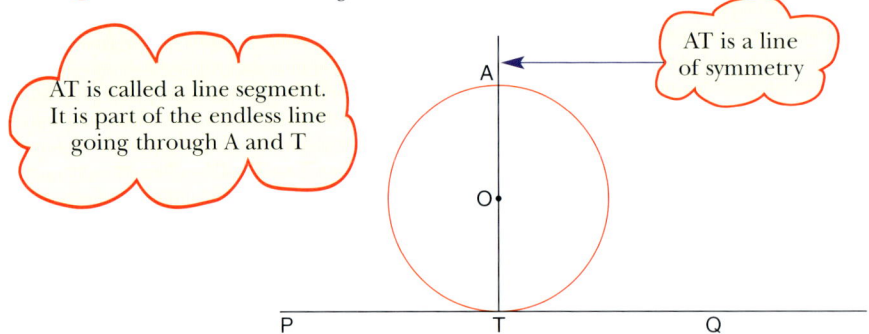

*AT is called a line segment. It is part of the endless line going through A and T*

*AT is a line of symmetry*

? Look below. What size are angles OTP and OSP?

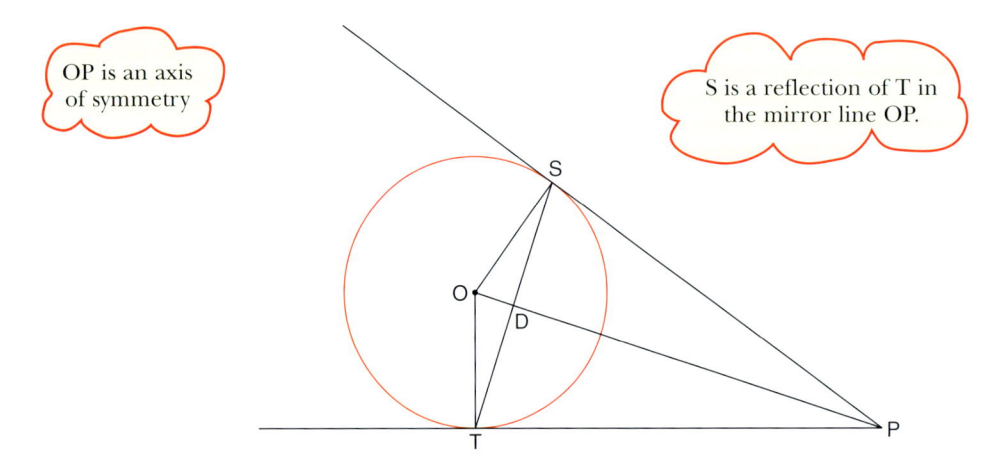

*OP is an axis of symmetry*

*S is a reflection of T in the mirror line OP.*

? Is SP equal to TP?

? Are tangents drawn from an external point always equal in length?

? What can you say about the angles at D?

? Is a line from the centre of a circle to the mid-point of a chord always perpendicular to the chord?

? Does the perpendicular bisector of a chord always pass through the centre of the circle?

# 9: Circles and tangents

**1** Find the angles marked with letters, giving reasons for your answers.

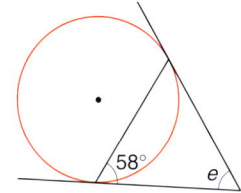

a) [diagram: tangent from external point, angle 54° at centre, angle a]

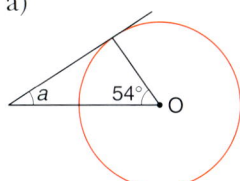

b) [diagram: angle b at centre O, 23° at external point]

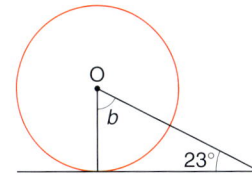

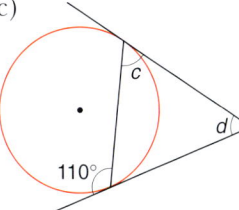

c) [diagram: 110° at centre, angles c and d]

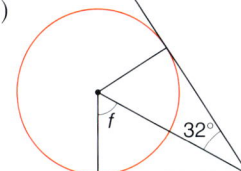

d) [diagram: 58° between tangent and chord, angle e]

e) [diagram: angle f at centre, 32° at external point]

f) [diagram: 70° at top, angle g]

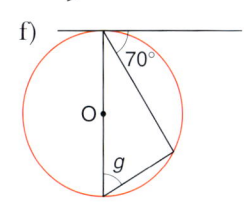

**2** Find the angles marked with letters. Give reasons for your answers.

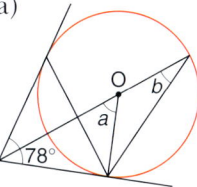

a) [diagram: angles a, b at O, 78° at external point]

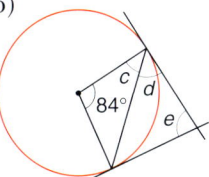

b) [diagram: angles c, d, e, 84° at centre]

c) [diagram: angle f at O, 66° at external point]

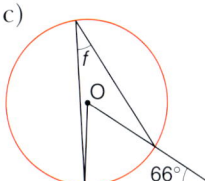

**3** The point O is the centre of a circle of radius 8 cm and OP is 17 cm. The tangents from P to the circle touch it at Q and R.

Draw a diagram and use Pythagoras' rule to calculate the length of PQ.

**4** The tangents from a point P to a circle are of length 35 cm. P is 37 cm from the centre of the circle.

Draw a diagram and calculate the radius of the circle.

**5** A circle, centre O, has radius of 10 cm. AB is a chord of the circle and AB = 12 cm. Work out the shortest distance from O to the chord AB.

## 9: Circles and tangents

# Finishing off

**Now that you have finished this chapter you should know that:**

- ★ opposite angles of a cyclic quadrilateral add up to 180°
- ★ the angle in a semi-circle is a right angle
- ★ angles in the same segment are equal
- ★ the angle at the centre is twice the angle at the circumference
- ★ the tangent is perpendicular to the diameter at the point of contact
- ★ tangents from a point to a circle are equal in length
- ★ the centre of a circle lies on the perpendicular bisector of any chord.

Use the questions in the next exercise to check that you understand everything.

## Mixed exercise

**1** Find the angles marked with letters, giving reasons for your answers.

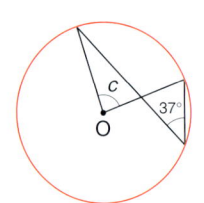

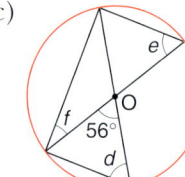

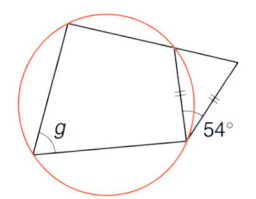

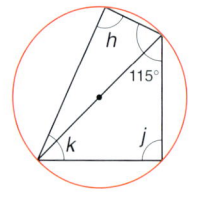

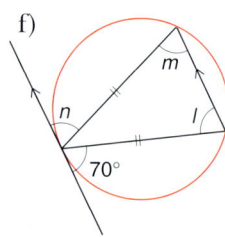

a) [circle with 48°, b, a]

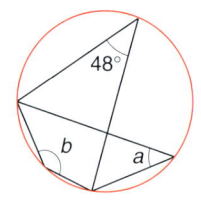

b) [circle with centre O, c, 37°]

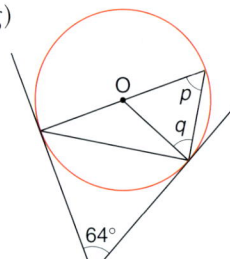

c) [circle with centre O, e, f, 56°, d]

d) [cyclic quadrilateral with g, 54°]

e) [circle with h, 115°, k, j]

f) [circle with tangent, m, n, l, 70°]

g) [circle with O, p, q, 64°]

h) [circle with r, 64°]

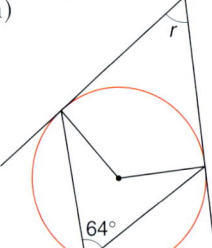

i) [circle with centre O, 120°, 28°, s]

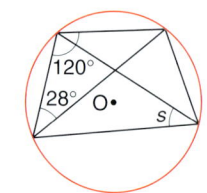

# 9: Circles and tangents

**Mixed exercise**

**2** In the diagram, O is the centre of the circle and angle ATB is 50°.

TA and TB are tangents.

a) Find angle AOB.

b) Find angle ACB, giving a reason for your answer.

**3** In the diagram, the circle through A, B, C, D and E has centre O, which lies on AC.

AT is the tangent to the circle at A.

a) Explain why angles TAC and ABC are right angles.

b) Angle TAB = 58°. Find the following angles, in each case giving a reason for your answer.

   (i) BAC        (ii) ACB

   (iii) ADB      (iv) AEB.

**4** O is the centre of a circle. BC is parallel to ED.

Calculate the unknown angles. Give reasons for your answers.

**5** PQ is 2 cm and OQ is 3 cm.

Find the lengths of PR and PS, the tangents from P to the circle.

**6** A circle, centre O and radius 9 cm, touches a circle, centre C and radius 4 cm, as shown.
The line AB is a tangent to both circles.

Find the length of AB.

(Hint: draw a line through C parallel to AB.)

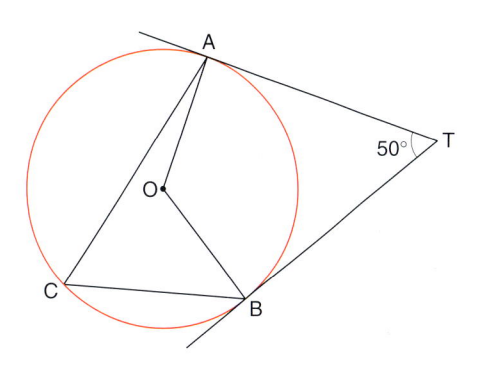

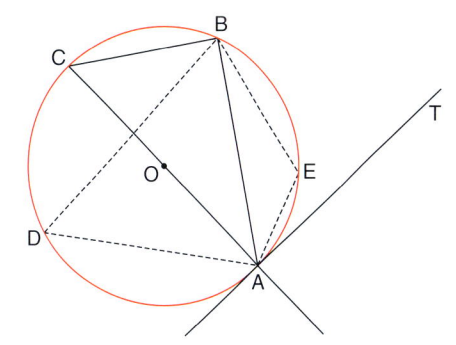

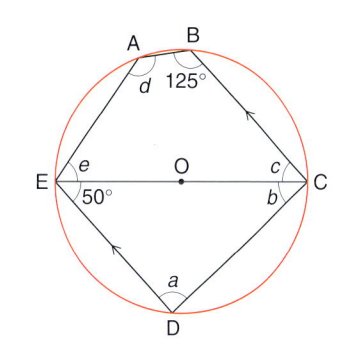

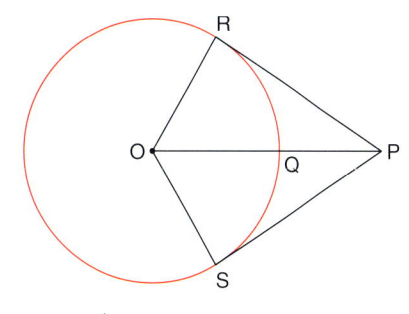

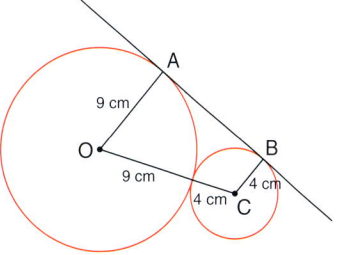

115

# Ten

# Manipulating expressions

**Before you start this chapter you should be able to**

★ expand brackets    ★ find the common factors of a set of numbers.

## Like terms

*Look at these expressions*
a) $8a + b$   b) $8x + x^2$   c) $8x^2 + 6x^2$   d) $8ab + 5ba$

*Can they be simplified?*

a)  The terms in $8a + b$ contain different unknowns, $a$ and $b$, so they are **unlike terms**. The expression cannot be simplified.

b)  The two terms in $8x + x^2$ contain the same unknown, $x$, but to different powers. Terms in $x$ and $x^2$ are also unlike terms, and so the expression cannot be simplified.

c)  The terms $8x^2$ and $6x^2$ are **like** terms. $8x^2 + 6x^2 = 14x^2$.

d)  The terms $8ab$ and $5ba$ are like, because $ab$ and $ba$ are the same ($a \times b = b \times a$). $8ab + 5ba = 13ab$ (or $13ba$).

You can add and subtract like terms to simplify an expression. But you cannot simplify an expression that has no like terms.

### Example
Simplify     $8a + 5a^2 + 5 + 8a^2 - a + 8 - 4a \times a$

### Solution
Collect like terms   $5 + 8 + 8a - a + 5a^2 + 8a^2 - 4a^2$   (This is $4a \times a$)
Tidy up              $13\quad + 7a\quad + 9a^2$

### Example
Simplify     $5x(1 + x) - 2x + x^2$   (All the terms here are unlike)

### Solution

$5x(1 + x) - 2x + x^2$   (Expand the brackets)
$= 5x + 5x^2 - 2x + x^2$   (Collect like terms)
$= 5x - 2x + 5x^2 + x^2$
$= 3x + 6x^2$   (Tidy up)

## 10: Manipulating expressions

**Exercise**

**1** State whether these pairs of expressions are identical (the same whatever the value of the letters).

a) $zx$ and $xz$    b) $2 \times t \times t$ and $2t^2$    c) $2$ and $2t^2$

d) $2x$ and $x^2$    e) $3(c+5)$ and $3c+5$    f) $7(1+k)$ and $7+7k$

**2** Separate each of these into two separate lists of like terms, then add together the like terms.

a) $2x^2, 4x, 3x^2, 7x$

b) $y^2, 3y, 2y^2, y, -y^2$

c) $3u^2, 4u, -2u, u, 6u^2, -2u$

d) $3p, -3p^2, 3, 6p, 5p^2, 2, 6p, -6$

**3** Copy each of these and simplify it by collecting together the like terms. Remember to work down your page.

a) $3x + 2y + 5x + y$    b) $6a + b + 4a + 9b$    c) $2p + 3q + p + q$

d) $5s + 7t + 5s + 2t$    e) $3l + 8m + 5m + 2l$    f) $3g + 2h - 3g + 3h$

g) $x + 5 + 3x - 2$    h) $5x - 2y - 3 - 3x + 6$

i) $4a - 3b - 3c + 2a + 3b - 3c$    j) $3p - 2q + 5p + 2q + 6 + 5$

**4** Copy each of these and simplify it by collecting together the like terms.

a) $x^2 + 3x - x + 1$    b) $x^2 - 4x + 2x - 3$    c) $y^2 - 2y - y + 2$

d) $2y^2 - 3y + 7 + y$    e) $5a^2 + 2a + 3a^2 - 3a$    f) $7 - 2x - 3x + x^2$

g) $21 + 6d - 8d - 21$    h) $10x^2 + 6 - 3x - 4 + 2x$    i) $9x - x^2 + 3x - 4x^2$

j) $t^2 + 3t - 3t + 9$

**5** Expand the brackets and collect the like terms.

a) $m(3 + m) + m^2$    b) $k + 2k(1+ k)$    c) $4x(5 + x) - 2x^2$

d) $y^2 + y(4 - y)$    e) $x(x + 2) - 4(x + 2)$    f) $a(a + 2) + 2(a + 2)$

g) $h(1 + h) + 5(1- h)$    h) $h(h - 3) - 6(h - 3)$    i) $x^2(1 + x) - 3x^3$

### Investigation

Show that $1 + 2 + 2^2 + 2^3 + 2^4 + 2^5 + 2^6 = 2^7 - 1$

How can you generalise this result?

## 10: Manipulating expressions

# Factorising

**?** *What are the common factors of 12 and 16?*

*What are the common factors of $12a$ and $16a^2$?*

The expression $3x + 12$ has two terms.
3 is a common factor.

> Divide each term by the common factor, and put the factor outside the bracket.

You can write the expression as $3(x + 4)$.

Writing $3x + 12$ as $3(x + 4)$ is called **factorising** the expression.
Factorising is just the reverse of expanding the brackets.

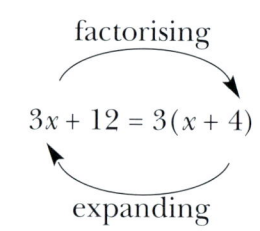

$$3x + 12 = 3(x + 4)$$
(factorising / expanding)

### Example

Factorise  a) $2x + 4y - 6$  b) $15x^2 - 3x$

### Solution

a)
$$2x + 4y - 6$$
$$= 2(\quad\quad)$$
$$= 2(x + 2y - 3)$$

> The number 2 is a common factor. Write it outside a bracket.

> Decide what goes inside. Remember to divide each term by 2.

Check: Expand the bracket:  $2 \times x + 2 \times 2y - 2 \times 3 = 2x + 4y - 6$ ✓

b)
$$15x^2 - 3x$$
$$= 3(5x^2 - x)$$
$$= 3x(5x - 1)$$

> The number 3 is a factor.

> The unknown, $x$, is also a factor

> You cannot factorise this any further

Check: Expand the bracket: $3x \times 5x - 3x \times 1 = 15x^2 - 3x$ ✓

### Example

Simplify $2(2x + 5y) + 3(x + 6y)$, giving your answer in factorised form.

### Solution

$$2(2x + 5y) + 3(x + 6y)$$

Expand brackets  $\quad 4x + 10y + 3x + 18y$

Collect like terms  $\quad 4x + 3x + 10y + 18y$

(Tidy up)  $\quad\quad\quad 7x + 28y$

Factorise  $\quad\quad\quad 7(x + 4y)$

## 10: Manipulating expressions

**1** Write down the common factors of each set of terms.

a) 12, 4x   b) 3x, 12y, 9   c) $7a^2$, 5a   d) $25b^2$, 15b

**2** Decide on the common factor in each of these sets of terms.

Write each term as the common factor × something else.

a) 12, 3c   b) 16, 8y   c) $7n^2$, n

In these there is more than one common factor.

d) $5x^2$, 10x   e) 14y, $7y^2$   f) $4x^2$, 6x

**3** Factorise these expressions. Remember to copy out each one and work down your page.

Check your answers by expanding the brackets.

a) 4t + 8   b) 6 – 3m   c) 2 + 18b
d) 5x + 10   e) 9z – 33   f) 5x – 5
g) 2a + 6b + 4c   h) 3x – 3y + 9z   i) 4a – 22b + 6c
j) 8p + 6q – 4r   k) 14l – 7m – 49n   l) 16a + 24b – 32c

**4** Factorise these expressions and check your answers.

a) $2x + x^2$   b) $3y - 2y^2$   c) $5x^2 - 4x$
d) $10y^2 + 7y$   e) $3x^2 + 12x + 6$   f) $21 + 7x - 14x^2$
g) $8g^2 - 16g - 4$   h) $15 + 5x + 20x^2$   i) $6y^2 - 12yx$
j) $3xy^2 - 11y^2$   k) $4st + 8t$   l) $3ab - 6ac + 9ad$

**5** For each of the following expressions

(i) expand the brackets
(ii) simplify it by collecting like terms
(iii) factorise the answer to (ii) if possible.

a) 3(x + 2y) + 5(x + 6y)   b) 4(2x + 3) + 3(x + 7)
c) 4(3p + 2q) + 3(p + 4q)   d) 5(x + y) + 3(x – 2y) + 2(x + 3y)
e) 6(3a – b + 2c) + 2(a + 3b + 2c)   f) 2(l + m + n) + 3(l – m + n) + m
g) 3(x + 2y) – 2(x + 4y) + 5(x + 4y)   h) 12(3p + 2q) – 4(2p + q) – 3(7p + 2q)
i) 5(x – 2y – 3z) – 3(x – 2y – 5z)   j) –2(r – s) – 4(2r – 3s) + 12(r – s)
k) x(x – 3) + 3(x + 3)   l) x(x – 4) – 4(x – 4) –16

## 10: Manipulating expressions

# Expanding two brackets

You have already expanded expressions like $2(x+3)$ and $x(x+3)$. You know that you have to multiply each term inside the bracket by the term outside it.

How would you do $(x+2)(x+3)$?

Start like this $\qquad x(x+3) + 2(x+3)$

$\qquad\qquad\qquad x^2 + 3x + 2x + 6$

Now collect like terms $\qquad x^2 + 5x + 6$

**?** *Check your answer by putting in $x = 10$ and working out $12 \times 13$*

You need to multiply each number in the first bracket by each number in the second.

One order for doing this is **F**irsts **O**utsides **I**nsides **L**asts (remember it as **FOIL**).

If you join the numbers as you multiply, the lines look like a smiley face.

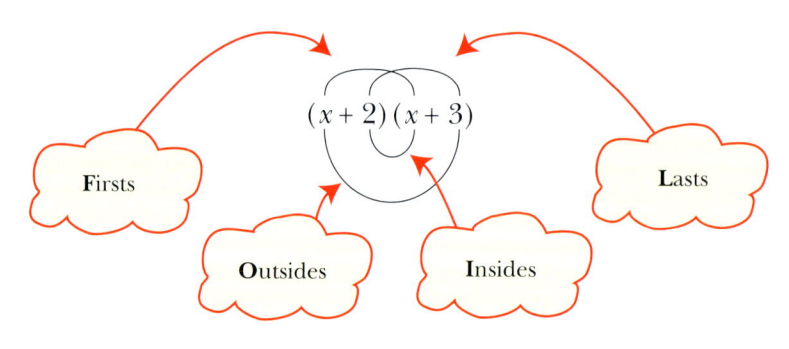

$(x+2)(x+3)$

Firsts  Outsides  Insides  Lasts

$x^2 + 3x + 2x + 6$

$x^2 + 5x + 6.$

**?** Use this method to work out $102 \times 104$ without a calculator.

### Example

Expand $(2x-5)(x-3)$.

### Solution

F  O  I  L

$2x \times x + 2x \times (-3) + (-5) \times x + (-5) \times (-3)$

$2x^2 - 6x - 5x + 15$

$2x^2 - 11x + 15$

## 10: Manipulating expressions

**1** Write these as briefly as possible.

a) $2a \times 3$  b) $2c \times 3c$  c) $4y \times 5y$

d) $-3 \times 2x$  e) $3 \times (-4x)$  f) $-3x \times (-5)$

**2** Expand and then simplify each of these.

a) $(x+1)(x+3)$  b) $(y+2)(y+5)$  c) $(4+x)(3+x)$

d) $(5+y)(6+y)$  e) $(x-1)(x+2)$  f) $(y+3)(y-2)$

g) $(x-6)(x+2)$  h) $(y+1)(y-5)$  i) $(x-7)(x-2)$

j) $(y-3)(y-4)$  k) $(x-3)(x-5)$  l) $(y-6)(y-5)$

**3** a) Calculate $21^2$ by expanding $(20+1)(20+1)$.

b) Calculate $31^2$ by the same method.

c) Expand $(x+1)^2 = (x+1)(x+1)$.

d) Use your answers to part c) to find $41^2$ without a calculator.

**4** Expand each of these and then simplify your answer.

a) $(a+3)(a+3)$  b) $(x+5)^2$  c) $(a-3)(a-3)$

d) $(x-5)^2$  e) $(2y+1)^2$  f) $(3x+2)^2$

**5** Expand and then simplify these.

a) $(2x+5)(3x+2)$  b) $(2x-5)(3x-2)$

c) $(2x+5)(3x-2)$  d) $(2x-5)(3x+2)$

**6** Expand each of these into 4 terms.

a) $(x+1)(y+1)$  b) $(a+3)(d+2)$  c) $(x-1)(y-1)$

d) $(c+4)(k-10)$  e) $(a-3)(x+4)$  f) $(x+10)(y-1)$

**7** The rectangle in the diagram has sides of length $(x+1)$ and $(x+2)$.

a) Write down an expression for its area.

b) Write down the area of each smaller part: A, B, C and D.

c) Expand $(x+1)(x+2)$.

d) Which rectangles correspond to the middle term of c)?

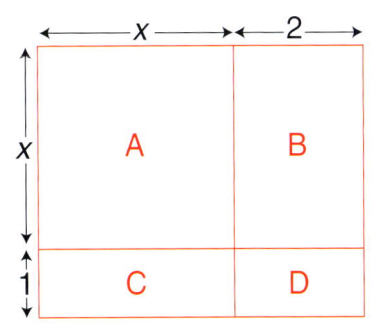

## 10: Manipulating expressions

# Squares

### Squaring brackets

You can write $(a + b)(a + b)$ as $(a + b)^2$.

Look at this square.

Its sides have length $(a + b)$,

so its area is $(a + b)^2$.

It is made up of 4 rectangles.
When you add their areas you get

$$a^2 + ab + ba + b^2 = a^2 + 2ab + b^2$$

It is important to remember that $(a + b)^2$ is *not* just $a^2 + b^2$.
There is a middle term $+2ab$.

Forgetting the $+2ab$ is like forgetting to include the two blue rectangles!

Expand $(a - b)(a - b)$. Be careful with the signs.

Check your answer by substituting $a = 10$ and $b = 1$.

### The difference of two squares

When you multiply $(a + b) \times (a - b)$
you get $a^2 - b^2$.

This is a useful result. It is called
the **difference of two squares**.

It is often written

$$a^2 - b^2 = (a + b)(a - b)$$

so when $a = 10$ and $b = 1$

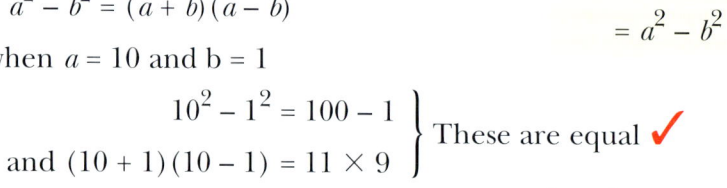

$10^2 - 1^2 = 100 - 1$
and $(10 + 1)(10 - 1) = 11 \times 9$ } These are equal ✓

$a^2 - ab + ba - b^2$
$= a^2 - b^2$

Use some other values of a and b to check that $a^2 - b^2 = (a + b)(a - b)$.

Work out $143^2 - 142^2$ without using a calculator.

On this page you have met three important results.

$(a + b)^2 = a^2 + 2ab + b^2$
$(a - b)^2 = a^2 - 2ab + b^2$
$(a + b)(a - b) = a^2 - b^2$

These are all worth remembering.
They are true when you replace *a* and *b* with any letters or numbers.

# 10: Manipulating expressions

**Exercise**

**1** a) Choose a number for $x$ and draw a square with sides $x$ units on squared paper.
   b) Work out $2x$ and draw another square with sides $2x$.
   c) How many of your first square will fit into your second?
   d) What is another way to write $(2x)^2$?

**2** Simplify each of these.
   a) $3x \times 3x$    b) $5y \times 5y$    c) $(3x)^2$
   d) $(5y)^2$    e) $(4a)^2$    f) $(10n)^2$

**3** Expand these brackets.
   a) $(x+4)(x+4)$    b) $(y+3)^2$    c) $(c-3)^2$
   d) $(n-5)^2$    e) $(2x+5)(2x+5)$    f) $(3+2t)^2$
   g) $(2y-3)^2$    h) $(5d-1)^2$    i) $(x-4)(x+4)$
   j) $(y+3)(y-3)$    k) $(2x-5)(2x+5)$    l) $(3+2t)(3-2t)$

**4** Use the results on the opposite page to write down the answers to these.
   a) $(x+y)^2$    b) $(x-y)^2$    c) $(x+7)^2$
   d) $(x-7)^2$    e) $(2x+3)^2$    f) $(4n-3)^2$
   g) $(5+z)^2$    h) $(3-x)^2$    i) $(2x+3y)^2$
   j) $(2x-3y)^2$    k) $(x+y)(x-y)$    l) $(x+7)(x-7)$
   m) $(2x+1)(2x-1)$    n) $(4n-9)(4n+9)$    o) $(5+z)(5-z)$
   p) $(3p-q)(3p+q)$    q) $(2x+3y)(2x-3y)$    r) $(10a-2b)(10a+2b)$

**5** a) Write $19 \times 21$ using the numbers 20 and 1.
   b) Explain how you could calculate $19 \times 21$ using two squares.
   c) Work out $29 \times 31$ in a similar way.

---

Draw two squares on squared paper as shown (you choose $a$ and $b$). Then cut out 2 rectangles each with sides $a$ and $b$.

Cover your squares with your two rectangles so that you are left with a small square with sides $(a-b)$.

Draw a diagram to show what you have done; mark on it the separate areas. What does this illustrate?

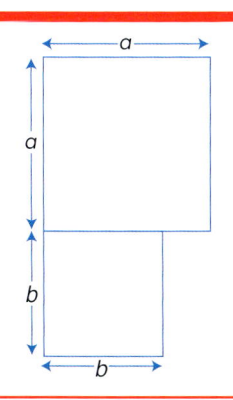

## 10: Manipulating expressions

# *Finishing off*

**Now that you have finished this chapter you should be able to**

★ factorise expressions by taking out a common factor
★ expand two brackets and simplify the result
★ square a bracket
★ recall that $a^2 - b^2 = (a+b)(a-b)$.

Use the questions in the next exercise to check that you understand everything.

### Mixed exercise

**1** Factorise each of these.
   a) $20s + 15$
   b) $26z - 13y$
   c) $18d + 6c - 3b$
   d) $2x + x^2$
   e) $t^2 - 12t$
   f) $5x^2 - 15x$

**2** In each of these, expand the brackets and collect like terms. If possible, factorise the answer.
   a) $2(3x + y) + 4(x + 2y)$
   b) $5(3p + 7) - 2(4p + 7)$
   c) $3(2a - 5b) - 5(a - 3b)$
   d) $k + 2k(1 - k)$
   e) $4x(3 + 2x) - 7x^2$
   f) $a(a + 6) - 5(a + 12)$

**3** Expand each of these and simplify the result.
   a) $(x + 12)(x + 11)$
   b) $(x + 7)(x - 5)$
   c) $(x - 9)(x + 6)$
   d) $(x - 11)(x + 3)$
   e) $(x - 20)(x - 1)$
   f) $(x - 13)(x - 3)$
   g) $(2x + 5)(x + 1)$
   h) $(3x + 4)(5x - 6)$
   i) $(a + 3)(4a - 3)$

**4** Write each of these without brackets.
   a) $(7b)^2$
   b) $(3t)^2$
   c) $(3x)^3$
   d) $(2x^2)^2$
   e) $(2x^2)^3$

**5** Expand each of these.
   a) $(x + 1)^2$
   b) $(y + 4)^2$
   c) $(t - 3)^2$
   d) $(b - 5)^2$
   e) $(2x + 1)^2$
   f) $(3x - 2)^2$
   g) $(6y - 1)^2$
   h) $(x^2 + 1)^2$
   i) $(1 + x)^2$

# 10: Manipulating expressions

**Mixed exercise**

**6** Expand each of these.
  a) $(x+y)^2$
  b) $(x+2y)^2$
  c) $(x+y)(x+2y)$
  d) $(x-y)(x+2y)$
  e) $(2a+2b)^2$
  f) $(2(a+b))^2$

**7** Expand each of these.
  a) $(x-1)(x+1)$   b) $(a-5)(a+5)$
  c) $(n+7)(n-7)$   d) $(2x+1)(2x-1)$
  e) $(3t-4)(3t+4)$  f) $(x+2y)(x-2y)$

**8** Pythagoras' rule states that in the right-angled triangle,
  $a^2 + b^2 = c^2$
  This can also be written as $a^2 = c^2 - b^2$.
  a) Show that $(c-b)(c+b) = c^2 - b^2$
  b) When $c = 13$ and $b = 12$, show that $a^2 = 25$. What is $a$ in this case?
  c) Find $a$ when $c = 25$ and $b = 24$.

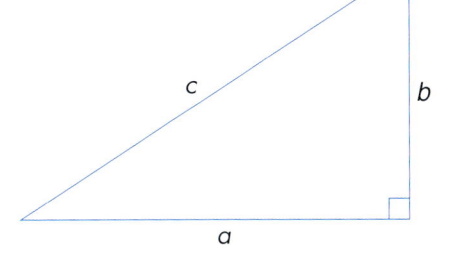

**9** Using the fact that $(a-b)(a+b) = a^2 - b^2$, work these out.
  a) $17^2 - 7^2$  b) $7.6^2 - 2.4^2$

**10** Using the fact that $(a+b)^2 = a^2 + 2ab + b^2$, work these out.
  a) $31^2$  b) $41^2$

**11** Using the fact that $(a-b)^2 = a^2 - 2ab + b^2$, work these out.
  a) $19^2$  b) $29^2$

**12** a) Expand $(x+1)(y+1)$.
  b) Substitute $x = 20$ and $y = 30$ into your answer and check it is the same as $21 \times 31$.
  c) Use your answer to a) to calculate $31 \times 61$.

**13** a) Expand $(x+1)(x-1)$.
  b) Use your answer to part a) to calculate $21 \times 19$ and $51 \times 49$.
  c) Use a similar method to calculate $22 \times 18$ and $48 \times 52$.

# Eleven

# Probability

**Before you start this chapter you should**

★ know that probability can be represented on a scale from 0 to 1

★ be able to estimate probabilities from past data

★ be able to calculate the probability of a particular outcome from theory.

Use the following questions to check you still remember these topics.

## Reminder

- P(A) means the probability of event A occurring.
- P(not A) = 1 − P(A)
- Probabilities can be expressed as fractions or decimals – sometimes fractions are more appropriate, sometimes decimals are easier.
- Probabilities should *not* be expressed as ratios or as 1 in *n*.

**Revision exercise**

**1** Draw a probability scale. Mark on your scale the following points, and for each one say whether its position on the scale is exact or approximate. The probability that

  A  a tossed coin will land showing tails

  B  you will break the world land speed record on your bicycle

  C  you will take at least one breath in the next 5 minutes

  D  it will rain tomorrow

  E  you will dig up a valuable old coin in your garden

  F  there is a point marked **F** on your scale.

**2** Fran has planted 3 hyacinth bulbs in separate pots, to give to relations at Christmas. One of the bulbs is pink, one is blue and one is white. She cannot remember which is which, and the flowers are not yet showing.

Fran picks two at random to give to her grandmothers.

a) Estimate the probability that one is blue and the other is white.

b) Estimate the probability that one is white.

# 11: Probability

**3** A driving school prepares learners for the driving test and only allows them to take the test when they are competent, and should pass. However, some people do still fail and the school believes that this happens at random.

Of the 125 people who started learning with the school in January, 100 passed first time, 20 passed second time and 4 passed third time.

Estimate the probability that a learner from this driving school will

a) pass at the first attempt
b) pass at the second attempt
c) pass at the third attempt
d) fail three driving tests.

**4** Craig is an amateur astronomer.

He keeps this 'shooting star' diary one April, recording a tick each night that he sees one, a cloud each night the sky is obscured by cloud, and a cross each clear night that he does not see a shooting star.

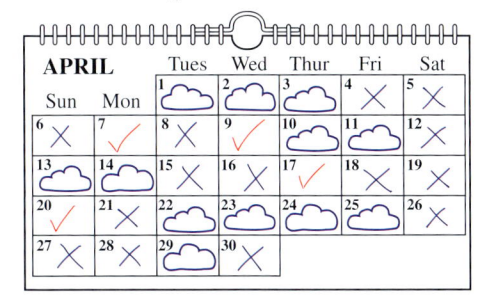

APRIL calendar:
- Sun 6 ✗, 13 ☁, 20 ✓, 27 ✗
- Mon 7 ✓, 14 ☁, 21 ✗, 28 ✗
- Tues 1 ☁, 8 ✗, 15 ✗, 22 ☁, 29 ☁
- Wed 2 ☁, 9 ☁, 16 ✗, 23 ☁, 30 ✗
- Thur 3 ☁, 10 ☁, 17 ✓, 24 ☁
- Fri 4 ✗, 11 ☁, 18 ✗, 25 ☁
- Sat 5 ✗, 12 ✗, 19 ✗, 26 ✗

a) Use the diary data to predict the probability that Craig will see a shooting star on any night next April.

b) Estimate the probability that the sky is hidden by cloud on any night.

c) On how many of the cloudy nights do you think there would have been shooting stars? Explain your reasoning.

d) Do you think that Craig has collected enough data to make confident estimates of these probabilities?

**5** You are playing Monopoly. You throw a red die and a blue die. Your score is the sum of the two numbers.

a) Copy and complete this table showing the possible outcomes.

b) From your table work out the probability of scoring

(i) 7, which means you go to jail

(ii) 2 or 12, which means you can buy a station

(iii) 11, which means you pay a lot of rent

(iv) 10, which means you take a card from Community Chest

(v) a double, which means you get a second throw.

Red die score / Blue die score table:

|   | 1 | 2 | 3 | 4 | 5 | 6 |
|---|---|---|---|---|---|---|
| 1 | 2 |   | 4 |   |   |   |
| 2 |   |   |   |   |   |   |
| 3 |   |   |   |   |   |   |
| 4 |   |   |   |   |   |   |
| 5 |   |   |   |   |   |   |
| 6 |   | 8 |   |   |   |   |

# 11: Probability

## Two outcomes: 'either, or'

Avonford Car Auctions holds a monthly Company Draw. One employee is picked (at random) by the company computer to win £200.

This week several people are off work because of a flu epidemic, and some people are away on courses.

What is the probability that the employee who wins the draw is absent, either with flu or on a course?

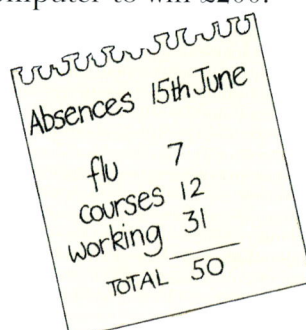

*Absences 15th June*
*flu 7*
*courses 12*
*working 31*
*TOTAL 50*

You can see that there are 50 employees in total, so there are 50 possible winners. Of those 50:

7 have flu: $\quad P(\text{flu}) = \dfrac{7}{50}$

12 are away on a course: $\quad P(\text{course}) = \dfrac{12}{50}$

*Nobody with flu is on the course. They are all in bed.*

Of the 50 employees, a total of 19 are away either with flu or on a course, so you can write

$$P(\text{flu or course}) = \dfrac{19}{50}$$

Notice that in this case,

$$P(\text{flu or course}) = P(\text{flu}) + P(\text{course})$$

*To find the probability that the winner either has flu or is on a course you can add the probability that the winner has flu and the probability that the winner is on a course*

Of the 50 employees, 31 are in the sports club and 10 are in the fell-walking club. What is the probability that the winner is in one of these clubs?

You may have realised that to answer the question you need more information. Some of the fell-walkers may also be in the sports club: the outcomes are not **mutually exclusive**.

In fact, Ted, Liz and Nina are in both clubs. You want to add the probabilities without counting these people twice, so you can write

$$P(\text{sports or fell}) = P(\text{sports}) + P(\text{fell but not sports}) = \dfrac{31}{50} + \dfrac{7}{50} = \dfrac{38}{50}$$

*There are 10 people in the fell-walking club but 3 have already been counted in the sports club. 10 − 3 = 7*

> You can add the probabilities in 'either, or' situations, provided you make sure that the outcomes do not overlap.

**?** *How many people, altogether, are involved in the sports club or the fell-walking club?*

## 11: Probability

**1** When you select a card at random from an ordinary pack, what is the probability that it is either

a) a king or a queen?   b) a king or a heart?

**2** Dick plants mixed crocuses: 5 purple, 7 yellow and 8 white.

Work out the probability that the first one to flower in the Spring will be

a) purple   b) yellow or white   c) purple or white

d) neither purple nor white.

**3** Paul is a keen bird-watcher. One June he keeps a close watch on 65 nests of house-martins. He records the number of hatchlings in each one that survive and fly the nest. This is his table of the results.

| Number of hatchlings flying the nest | 0 | 1 | 2 | 3 | 4 | 5 | 6 |
|---|---|---|---|---|---|---|---|
| Frequency (number of nests) | 1 | 5 | 12 | 18 | 24 | 3 | 2 |

Using Paul's data, estimate the probability that a nest of house-martins will produce

a) 5 hatchlings that fly the nest

b) at least 4 hatchlings that fly the nest

c) fewer than 3 hatchlings that fly the nest.

**4** Make a table to show all the possible scores when two fair dice (one red, one blue) are thrown.

Find the probability that

a) the scores are the same (it is a double)

b) the score on the red die is greater than the score on the blue one

c) both dice show odd numbers

d) the total score is more than 7.

---

You are going to check whether real dice behave as your answers to question 4b) suggest. Find two dice of different colours. Draw a table like the one you drew for question 4 to record your results.

The relative frequency each time is the number of times red has scored more than blue, divided by the total number of throws so far.

You would expect the relative frequency to get closer and closer to your calculated probability. Is this noticeable after 10 throws? Extend the table, and see if it is closer after 20 throws, or 50 throws.

## 11: Probability

# Two outcomes: 'first, then'

When you select a playing card at random from an ordinary pack, the probability of selecting an ace is $\frac{4}{52}$. (There are 4 aces in the pack, and 52 cards altogether.) When you select one card, then replace it, then select another, the probability of an ace is $\frac{4}{52}$ each time.

What is the probability that both cards are aces?

In this situation, when you want to work out the probability of first one outcome, then another happening, you multiply the probabilities.

$$\frac{4}{52} \times \frac{4}{52} = \frac{1}{13} \times \frac{1}{13} = \frac{1}{169}$$

**The probability that both cards are aces is $\frac{1}{169}$**

**?** *What is the probability that both cards are black?*

Sometimes two events happen at the same time, but you can still think of them as first one, then another.

### Example

Calum goes into a shop to buy 2 light bulbs. He cannot tell from the packaging, but 6 of the 25 light bulbs on the shelf are faulty.

What is the probability that he picks up 2 faulty bulbs?

### Solution

When he chooses his first bulb, the probability of a faulty one is $\frac{6}{25}$.

When he chooses his second bulb, the probability of a faulty one is different. There are 24 bulbs left, of which 5 are faulty.

The probability of a faulty one is $\frac{5}{24}$.

The probability that Calum picks two faulty bulbs is $\frac{6}{25} \times \frac{5}{24} = \frac{1}{20}$.

*These events are not independent; the outcome of the first affects the probability of the second.*

**?** *What is the probability that Calum picks two good bulbs?*

*What is the probability that he picks one faulty and one good?*

*What combination is Calum most likely to pick?*

> **In 'first, then' situations, you multiply the probabilities, but you need to think carefully about those probabilities when the events are not independent.**

# 11: Probability

**1** Two coins are tossed. Find the probability that

   a) both land showing Heads

   b) one lands showing Heads, the other Tails.

**2** Angela, Basil and Connie travel to a meeting by train from different directions. Every train has a probability of 0.1 of being late, and you can treat them as independent events.

Find the probability that

   a) Connie is late

   b) Connie and Basil are both late

   c) Connie and Basil are late but Angela's train is on time

   d) all three trains are on time.

**3** Three coins are tossed. Find the probability that they land showing

   a) three Heads

   b) two Heads and a Tail (in that order)

   c) two Tails and a Head (in that order)

   d) two Tails and a Head (in any order)

   e) three Tails.

**4** Rowena does a survey of pedestrians in Avonford town centre. She asks them how they have travelled there, and the main purpose of their visit. She presents the results in these pie charts.

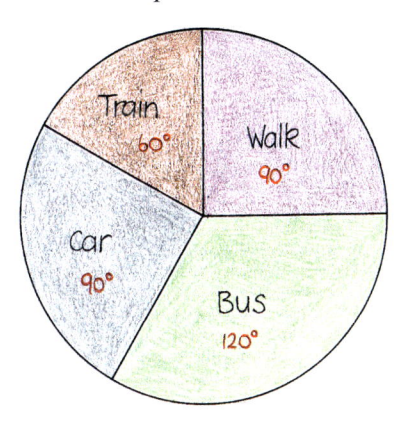

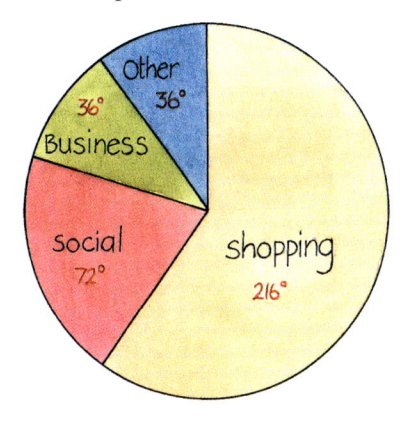

Pie chart 1: Train 60°, Walk 90°, Car 90°, Bus 120°

Pie chart 2: Other 36°, Business 36°, social 72°, shopping 216°

Using Rowena's pie charts, work out the probability that a pedestrian chosen at random

   a) has travelled into Avonford on foot

   b) is there mainly for a social event.

Assuming that the events are independent, what is the probability that a pedestrian chosen at random

   c) arrived on foot for a social event?

   d) arrived by train for business?

   e) arrived by bus or train for a shopping trip?

## 11: Probability

# Probability trees

The probability that a person (chosen at random) is left-handed is $\frac{1}{10}$. The probability that a person wears glasses is $\frac{1}{4}$.

*What is the probability that a person chosen at random is left-handed and wears glasses?*

*What is the probability that the person is right-handed and wears glasses?*

You can work out these probabilities using the ideas from the previous pages. Alternatively you can represent the situation in a **tree diagram**, which shows the probabilities of all possible outcomes.

Any person is taken to be either left-handed or right-handed: two alternatives. You show this as two branches of a 'tree', and write the probability on each branch.

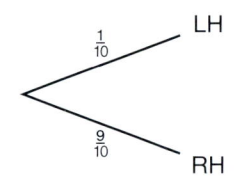

Any person either wears glass or doesn't. Again there are two alternatives. You extend the diagram like this.

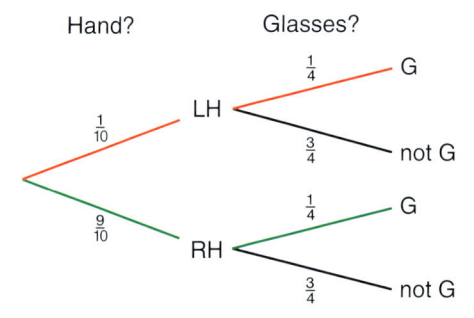

The probability that a person is left-handed and wears glasses is given by multiplying the probabilities along the red branches:

$$P(\text{LH and G}) = \frac{1}{10} \times \frac{1}{4} = \frac{1}{40}$$

The probability that a person is right-handed and wears glasses is given by

$$P(\text{RH and G}) = \frac{9}{10} \times \frac{1}{4} = \frac{9}{40} \text{ (the green branches)}$$

*There are four possible outcomes. The probabilities of two of them have been found above. Find the probabilities of the other two outcomes, and add all four probabilities together.*

*What does this tell you?*

Probability trees are particularly helpful when there are more than two outcomes from each event, and when the events are not independent.

# 11: Probability

**1** A student survey finds that 60% of students own a bicycle and 70% own a CD player. Assuming these to be independent, draw a probability tree and find the probability that a randomly chosen student owns

   a) a bicycle but no CD player

   b) a CD player but no bicycle

   c) neither a CD player nor a bicycle

   d) both a CD player and a bicycle.

**2** Fred claims to be able to tell the colour of a Smartie by taste alone. A test is organised with Fred blindfolded. He has to pick 30 Smarties out of a big bowl containing equal numbers of red, green and brown Smarties. He eats each one and says what colour he thinks it is.

   a) How many out of the 30 would you expect Fred to get right if

   (i) his claim is false and he is just guessing?

   (ii) his claim is true and he really can tell them apart?

   b) In fact Fred can tell the green ones but guesses between the red and brown.

   Copy and complete this tree diagram for the situation.

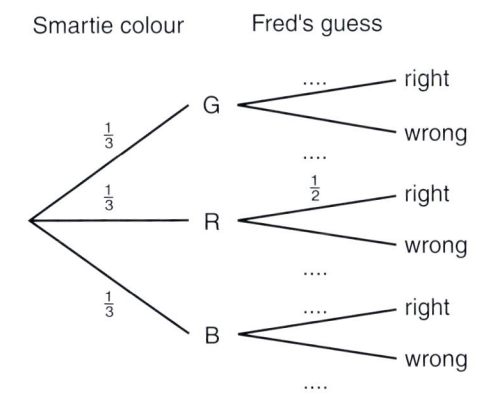

   c) What is the probability that Fred gets a Smartie right?

   d) In fact Fred gets 21 right. Is that about what you would expect?

---

Set up and carry out an experiment like the one in question 2. Write a short report on your findings, including tree diagrams, to show what the probabilities appear to be.

# 11: Probability

# Finishing off

**Now that you have finished this chapter you should be able to**

★ work out the probability of a particular outcome in an 'either, or' situation

★ work out the probability of a particular outcome in a 'first, then' situation

★ use a tree diagram to work out the probabilities of different outcomes for two or more events.

Use the following questions to check that you still remember these topics.

## Mixed exercise

**1** Alice draws a card from a pack of cards. Andrew draws a card from a different pack. Find the probability that both cards

a) are the same suit

b) are hearts

c) are the same number (or picture)

d) are court cards (jack, queen or king)

e) are the same colour

f) are *not* court cards.

**2** The probability that a person chosen at random can move his or her ears is 0.2, and the probability that the person can roll his or her tongue is 0.4.

a) Draw a probability tree to represent this situation.

b) What is the probability that a person, chosen at random, can move his or her ears *and* tongue-roll?

**3** Chris finds that in about 1 out of every 2 golf games he loses a ball. For Dennis, it is about 1 in every 3 games.

Estimate the probability that when they play each other

a) Chris loses a ball but Dennis doesn't

b) Dennis loses a ball but Chris doesn't

c) one (and only one) ball is lost

d) both lose a ball

e) neither loses a ball.

# 11: Probability

**Mixed exercise**

**4** The carnation plants at a garden centre have lost their labels and got mixed up. The manager knows that 40% are red, 20% are white, 30% are pink and 10% are yellow.

Amy buys two of the carnation plants. What is the probability that

a) they are both red?

b) they are both white?

c) they are both the same colour?

d) they are different colours?

e) one is red and one is white?

f) one is red and one is not?

**5** The probability that a new baby will be a boy is about 0.51.

a) Draw a tree diagram to show the probabilities for a woman who has two children.

b) What is the probability that she has two boys?

c) What is the probability that she has two girls?

d) What is the most likely combination?

**6** On Marie's route to work she drives through a junction with traffic lights. The whole traffic light sequence at this junction takes 2 minutes, and it runs as shown in the diagram.

- red — 45 seconds
- red and amber — 5 seconds
- green — 65 seconds
- amber — 5 seconds

What is the probability that on any journey Marie arrives at the junction when the lights are

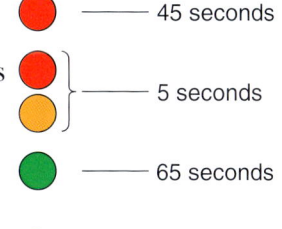

a) green?  b) red?  c) red and amber?

One day, Marie drives through the traffic lights 3 times. What is the probability that she arrives

d) at a green light every time?  e) at a red light every time?

---

Look up Gregor Mendel in an encyclopedia. Read about his experiments involving sweet-peas, and write a short description of them, explaining the importance of his findings.

Mendel's work is the basis for much of modern genetics.

# Twelve

# Locus

**Before you start this chapter you should be able to**

★ make accurate drawings, full size or to scale.

## Simple loci

The **locus** of a point means all the possible positions for that point.

Ainsley and Sarah are going on a camping holiday. They need to decide where to pitch their tent. Ainsley wants to be no more than 200 m from the shop. The map of the campsite shows the area where he would like to be.

All the points inside the circle are less than 200 m from the shop

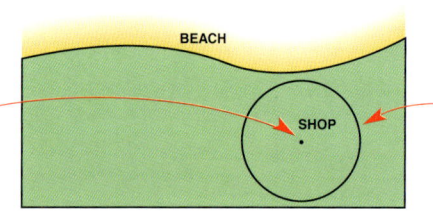

All the points on the circumference of the circle are exactly 200 m from the shop

*How could you describe the points outside the circle?*

**The locus of a point a fixed distance, $d$, from a fixed point, O, forms a circle, centre O and radius $d$.**

Sarah wants to be less than 100 m from the beach. The map shows the area where she would like to be.

All the points on this line are exactly 100 m from the beach

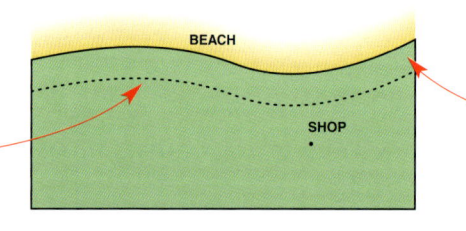

All the points in this area are less than 100 m from the beach

**The locus of a point a fixed distance, $d$, from a line, AB, forms a line parallel to AB and distance $d$ from AB.**

*Why is the line in this diagram dotted and not solid?*

All the points in the shaded area are both no more than 200 m from the shop and less than 100 m from the beach.

Ainsley and Sarah should camp somewhere in this area.

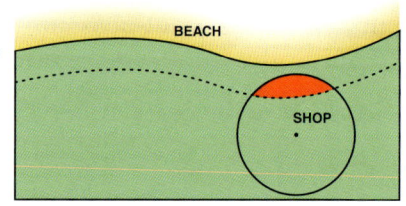

## 12: Locus

**1** For each part, draw this rectangle full size and shade in the required locus. Only shade points inside the rectangle.

a) The locus of all points less than 3 cm from A.

b) The locus of all points more than 2 cm from BC.

c) The locus of all points at least 1 cm from the centre of the rectangle.

d) The locus of all points more than 6 cm from C and more than 5 cm from D.

e) The locus of all points less than 3 cm from AB and at least 2 cm from B.

f) The locus of all points no more than 1 cm from the perimeter of the rectangle.

**2** A goat is tied to a point (marked G on the diagram) on the outside of a barn, 3 metres from the corner. The width of the barn is 5 metres. The rope is 5 metres long. Make a scale drawing of the diagram and shade in the locus of the points where the goat can go.

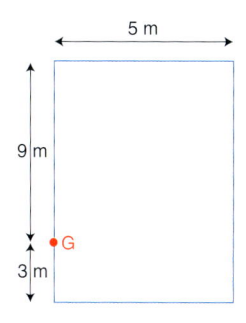

**3** Molly wants to plant a tree in her garden. She must not plant the tree within 2 metres of the house. Molly's house is 12 m long and 10 m wide. Make a scale drawing of the house and shade in the locus of the points where Molly must *not* plant the tree.

**4** Keith is looking for somewhere to live. He wants to be no more than 3 miles away from the station as he catches a train to work every morning. He is also a keen cinema-goer and would like to be no more than 5 miles away from the local cinema. The station and the cinema are 6.5 miles apart.

Make a scale drawing and show the area where Keith would like to live.

**5** Simon and his brother Mark are playing in a rubber dinghy in the sea. The coastguard has told them not to go more than 50 m from the shore. There is a rock 40 m from the beach and he has also told them not to go within 5 m of the rock.

Make a scale drawing and shade the area where they are allowed to go. (Assume that the shoreline is straight.)

Investigate the locus of a point on the circumference of a bicycle wheel as the bicycle moves along.

# 12: Locus

## A point equidistant from two fixed points

A new road is being built between two villages A and B. So that the new road makes as little disturbance as possible in the two villages, the road is being built so that it is always the same distance from A as from B.

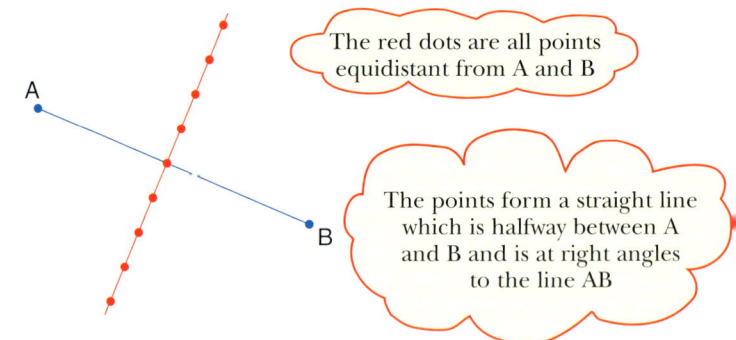

*The red dots are all points equidistant from A and B*

*The points form a straight line which is halfway between A and B and is at right angles to the line AB*

The position of the road will be the locus of a point **equidistant** (the same distance) from A and B.

This is called the **perpendicular bisector** of AB.

> **The locus of a point equidistant from two points is the perpendicular bisector of the two points.**

**?** *Where are the points which are nearer to A than to B?*

### Drawing a perpendicular bisector accurately

These instructions explain how to draw the perpendicular bisector of the line AB.

1. Place your compass point on A. Open the compass to a radius more than half the distance from A to B. Draw an arc each side of AB.

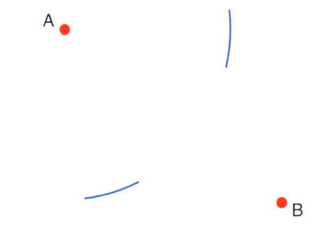

2. Leave the compass at the same radius. Put the compass point on B and make another arc each side of AB, so that they cross the other arcs.

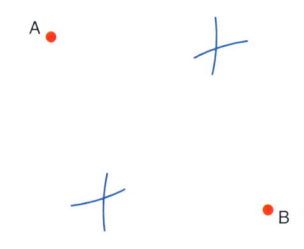

3. Draw a line joining the two intersections. This is the perpendicular bisector.

138

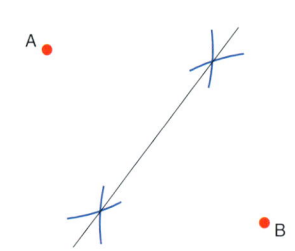

## 12: Locus

**Exercise**

**1**  a) Mark two points P and Q (not in a horizontal or vertical line). Construct the perpendicular bisector of PQ.

b) Mark 3 points on the perpendicular bisector. Measure the distance of each point from P and from Q to check that it is the same.

**2**  a) Draw a circle and mark three points, A, B and C on the circumference.

b) Construct the perpendicular bisectors of AB, BC and CA.

c) You should find that they meet at a point.

What is special about the point where they meet?

**3**  Sophie works in town S and her parents live in town T, 50 miles from town S. She is looking for a place to live somewhere between the two towns. She wants to be nearer to S than to T but would like to be within 30 miles of her parents.

Make a scale drawing and shade the area where Sophie would like to live.

**4**  The diagram below shows three schools A, B and C in a large town. Mr and Mrs Hammond and their son Michael are moving into the area, and Michael wants to go to school C. The Hammonds need to live somewhere which is nearer to school C than to school A, and nearer to school C than to school B.

Trace the diagram and shade the area where the Hammonds should live.

A
•


         •
         B

  C •

Find a map of your area with some of the local schools marked on it. Use perpendicular bisectors to show the areas which are nearer to each school than to any other school.

# 12: Locus

## A point equidistant from two lines

The designer of a new housing estate is putting two houses at the end of a cul-de-sac. She wants the boundary fence between the two houses to be the same distance from the wall of each house. The position of the fence will be the locus of a point equidistant from two lines.

*The red dots are all points which are the same distance from each house.*

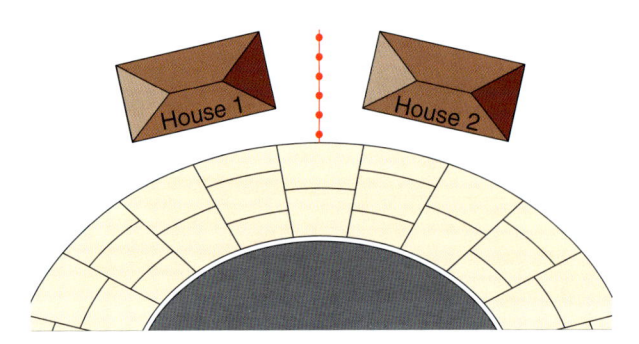

*The dots form a straight line which will be the position of the boundary fence.*

In this diagram, the lines showing the wall of each house have been extended until they meet, forming an angle. The red line showing the position of the fence has also been extended.

You can see that the fence line cuts the angle formed by the other two lines in half. It is called the **angle bisector**.

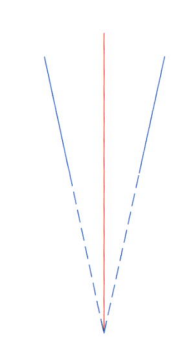

**The locus of a point equidistant from two lines is the angle bisector of the two lines.**

*Where are the points nearer to house 1 than to house 2?*

### Drawing an angle bisector accurately

1. Put the compass point on the point of the angle and mark off two points as shown.

2. Put the compass point on each of the points you have marked off and draw two arcs which meet each other.

3. Draw a line through the point where the arcs intersect to the point of the angle. This line is the angle bisector.

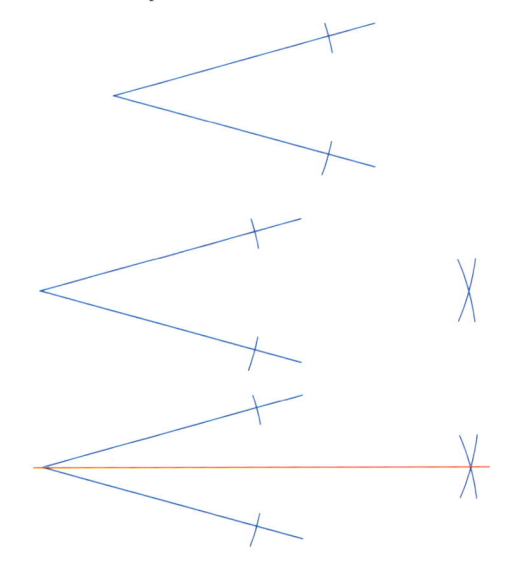

# 12: Locus

**1** a) Draw two lines to make an angle. Construct the angle bisector.

b) Measure the angle and check that the angle bisector cuts the angle in half.

**2** a) Draw a triangle. Construct the angle bisector of each of the three angles of the triangle.

b) The three angle bisectors should meet at a point. What is special about this point?

**3** Trace this triangle four times.

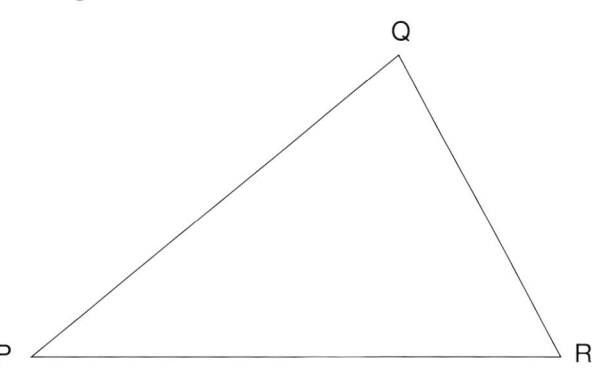

a) On the first diagram, shade the locus of all points nearer to PR than to PQ.

b) On the second diagram, shade the locus of all points nearer to QR than to PQ and within 3 cm of Q.

c) On the third diagram, shade the locus of all points nearer to PR than to QR and nearer to PQ than QR.

d) On the fourth diagram, shade the locus of all points nearer to PQ than to PR and nearer to Q than to P.

**4** This diagram shows a field in which there are some rabbits. The field is surrounded by hedges on sides AB, BC and CD. A fox appears at point F and all the rabbits run to the nearest hedge. Trace the field and divide it up, showing the areas which are nearest to each of the three hedges.

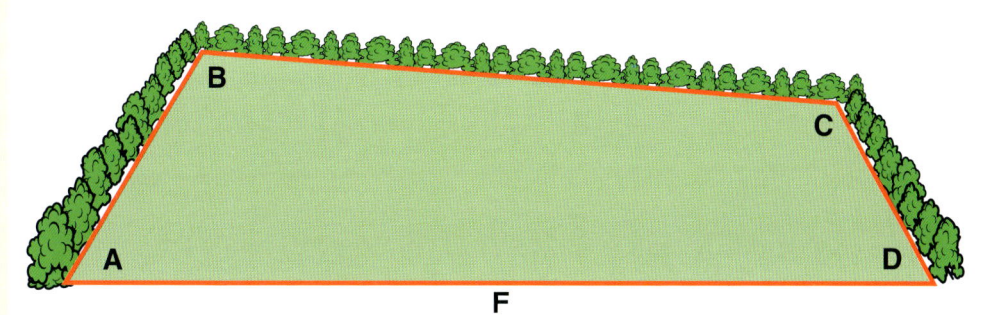

# 12: Locus

## Finishing off

### Now that you have finished this chapter you should know how to

★ construct the perpendicular bisector of a line

★ construct the bisector of an angle

★ solve problems involving loci, including intersecting loci

★ find the locus of a point which is a fixed distance from a point, a fixed distance from a line, equidistant from two points, or equidistant from two lines.

Use the questions in the next exercise to check that you understand everything.

**Mixed exercise**

**1** Mark a point X. Draw the locus of all points which are less than 4 cm from X.

**2** Mark points S and T 5 cm apart (not in a horizontal or vertical line). Draw the locus of all points equidistant from S and T.

**3** Draw an angle ABC of size 63°. Draw the locus of all points equidistant from the lines AB and BC.

**4** Draw a line XY 3 cm long. Draw the locus of all points exactly 2 cm from XY. (Hint: think carefully what happens to points at each end of the line.)

**5** Draw points M and N 8 cm apart (not in a horizontal or vertical line). Draw the locus of all points nearer to M than to N.

*For questions 6 to 11, trace the triangle XYZ and shade the locus of the point P. In all cases, P is inside the triangle XYZ.*

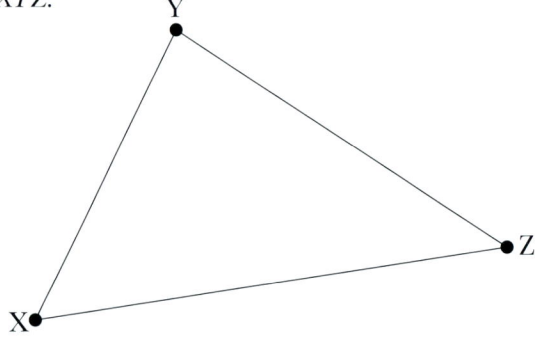

**6** P is less than 4 cm from X and less than 5 cm from Z.

**7** P is nearer to X than to Z and is no more than 2 cm from Y.

# 12: Locus

**8** P is less than 3 cm from XZ and less than 2 cm from Y.

**9** P is no more than 3 cm from X, less than 4 cm from Y and at least 5 cm from Z.

**10** P is nearer to XY than to XZ and is more than 1 cm from YZ.

**11** P is nearer to Y than to Z and is nearer to YZ than to XY.

**12** Jed is at a rock concert. He wants to be equidistant from the two speakers to get the best stereo effect. He also wants to be less than 10 m from the stage.

Make a scale drawing and show the possible places where Jed would like to be.

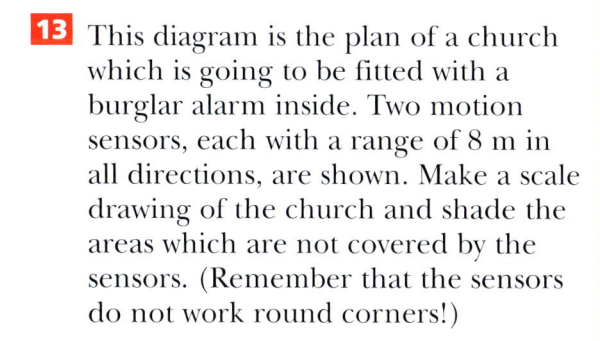

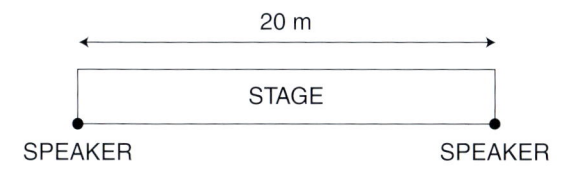

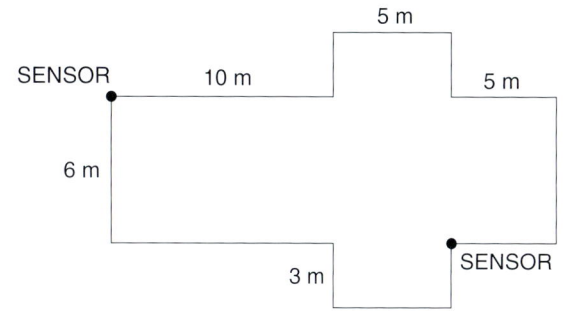

**13** This diagram is the plan of a church which is going to be fitted with a burglar alarm inside. Two motion sensors, each with a range of 8 m in all directions, are shown. Make a scale drawing of the church and shade the areas which are not covered by the sensors. (Remember that the sensors do not work round corners!)

**14** Alex needs to draw a line through A, perpendicular to the given line in these two cases.

He starts by drawing two arcs of a circle centre A.

Copy and complete the constructions.

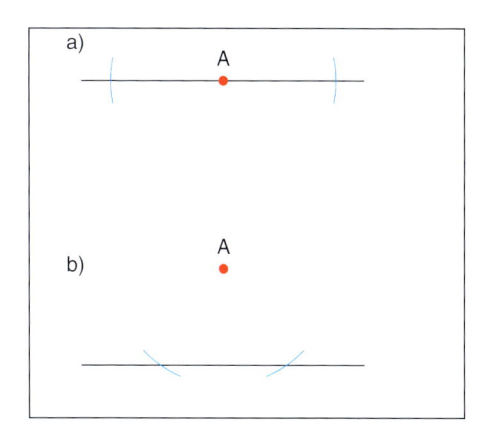

*Mixed exercise*

# Thirteen

# Quadratics

**Before you start this chapter you should**

★ be familiar with the work in Chapter 10.

## Factorising quadratic expressions

In this chapter you learn how to solve quadratic equations. The first step in this is to **factorise** a quadratic expression. This is the opposite of multiplying two brackets together (or expanding them).

The lines are the same, but written in opposite orders.

```
EXPANDING  (x+2)(x+3)                FACTORISING  x²+5x+6        3+2=5
(I usually don't                                                 3×2=6
write this line)  x(x+3) + 2(x+3)    Split the 5x  x²+3x+2x+6
I use FOIL        x² + 3x + 2x + 6   Factorise in pairs x(x+3)+2(x+3)
                  F   O    I    L
Tidy up           x²  +  5x  + 6     Finish off     (x+2)(x+3)  ← x+3 is
                                                                  in both
                                                                  brackets
```

Alternatively you can start by writing

$$x^2 + 5x + 6 = (x\ \ \ )(x\ \ \ )$$

*Now you need to work out the number term in each bracket*

Look for two numbers that multiply to give 6.

They could be $1 \times 6$ or $2 \times 3$.

Your two numbers must also add up to 5.

They are 2 and 3.

$$x^2 + 5x + 6 = (x + 2)(x + 3)$$

You must check by multiplying out the brackets.

```
CHECK
(x+2)(x+3)
x² + 3x + 2x + 6
x² + 5x + 6
```

### Example

Factorise $x^2 - 5x + 6$.

### Solution

In this case the numbers are –2 and –3.

$$x^2 - 5x + 6 = (x - 2)(x - 3)$$

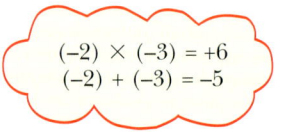

*(–2) × (–3) = +6*
*(–2) + (–3) = –5*

**?** *This solution is incomplete. What is missing?*

144

# 13: Quadratics

**Exercise**

**1** a) Find two numbers with product 5 and sum 6.
   b) Use these numbers to help you factorise $x^2 + 6x + 5$.
   c) Check your answer by expanding your brackets.

**2** a) Find two numbers with product 15 and sum 8.
   b) Use these numbers to help you factorise $y^2 + 8y + 15$.
   c) Check your answers by expanding your brackets.

**3** Factorise these and check your answers.
   a) $x^2 + 9x + 14$    b) $x^2 + 15x + 14$    c) $a^2 + 7a + 10$
   d) $a^2 + 11a + 10$   e) $n^2 + 5n + 4$      f) $t^2 + 4t + 4$
   g) $x^2 + 4x + 4$     h) $x^2 + 9x + 18$     i) $y^2 + 10y + 24$

**4** a) Find two negative numbers with product +5 and sum –6.
   b) Use these to help you factorise $x^2 - 6x + 5$.
   c) Check your answer by expanding your brackets.

**5** a) Find two numbers with product +15 and sum –8.
   b) Use these numbers to help you factorise $y^2 - 8y + 15$.
   c) Check your answer by expanding your brackets.

**6** Factorise these and check your answers.
   a) $x^2 - 3x + 2$     b) $x^2 - 4x + 3$      c) $x^2 - 6x + 5$
   d) $x^2 - 12x + 11$   e) $x^2 - 9x + 14$     f) $x^2 - 6x + 8$
   g) $p^2 - 9p + 18$    h) $a^2 - 15a + 44$    i) $x^2 - 45x + 44$

**7** Factorise these and check your answers.
   a) $y^2 - 2y + 1$     b) $a^2 + 13a + 22$    c) $x^2 - 10x + 21$
   d) $p^2 - 10p + 16$   e) $t^2 - 8t + 12$     f) $y^2 + 7y + 12$
   g) $a^2 + 2a + 1$     h) $x^2 + 6x + 9$      i) $y^2 - 4y + 4$
   j) $t^2 - 10t + 25$   k) $x^2 - 6x + 9$      l) $y^2 - 5y + 4$
   m) $p^2 - 12p + 36$   n) $x^2 + 14x + 49$    o) $x^2 - 25x + 100$

# 13: Quadratics

# More quadratic factorisation

In quadratics you have factorised so far the number terms have been positive (+). When you factorised these you got two brackets with the same signs inside, such as $(x-3)(x-2)$ or $(x+5)(x+2)$.

In the examples on this page the number term is negative (−), and you have to be a bit more careful with the signs.

Look at $x^2 - x - 20$.

*Note: −x means − 1x.*

You need to find two numbers that multiply to give −20 and add to give −1.

To multiply to get 20, you can have 20 and 1, 10 and 2 or 5 and 4.

Since it is −20, one of the numbers must be positive and one negative.

After trying the various pairs, you can see that the only one that works is −5 and +4.

The answer is $x^2 - x - 20 = (x-5)(x+4)$.

**?** *Check this by multiplying out the brackets.*

**?** *Copy and complete this table. Look carefully at the pattern of signs in your table.*

*What signs should you expect in the brackets when you factorise $x^2 - x - 12$?*

| Brackets | Quadratic |
|---|---|
| $(x + 3)(x + 5)$ | $x^2 + 8x + 15$ |
| $(x - 3)(x - 5)$ | |
| $(x - 3)(x + 5)$ | |
| $(x + 3)(x - 5)$ | |

In Chapter 10 you met quadratic expressions like $x^2 - 9$ (the difference of two squares). You can factorise $x^2 - 9$ by writing it as $x^2 - 0x - 9$.

**?** *Work this through to get the answer $(x+3)(x-3)$.*

*Use the same idea to factorise $x^2 - 4$.*

It is useful to remember that $x^2 - a^2$ factorises to $(x+a)(x-a)$.

When you need to factorise a quadratic expression whose terms have a common factor, you take out the common factor first.

**Example**
$$2x^2 + 6x + 4$$
$$= 2(x^2 + 3x + 2)$$
$$= 2(x+1)(x+2)$$

*The terms have a common factor 2.*

# 13: Quadratics

**1** a) Think of two numbers with product 6 and difference 5.
   b) Use these to help you factorise $x^2 + 5x - 6$.
   c) Factorise $x^2 - 5x - 6$.
   Check your answers by expanding the brackets.

**2** a) Think of two numbers with product 15 and difference 2.
   b) Use these to help you factorise $y^2 + 2y - 15$.
   c) Factorise $y^2 - 2y - 15$.
   Check your answers by expanding the brackets.

**3** Factorise these and check your answers.
   a) $x^2 + 10x - 11$  b) $x^2 - 10x - 11$  c) $x^2 + 6x - 7$
   d) $x^2 - 6x - 7$    e) $x^2 + 4x - 5$   f) $x^2 - 4x - 5$
   g) $x^2 + 5x - 14$   h) $x^2 - 5x - 14$  i) $x^2 + 3x - 88$

**4** Factorise these and check your answers.
   a) $a^2 + 7a - 18$   b) $a^2 - 7a - 18$  c) $y^2 + 9y - 10$
   d) $y^2 - 3y - 10$   e) $p^2 - 3p - 18$  f) $x^2 - x - 12$
   g) $x^2 + x - 20$    h) $a^2 + 8a - 20$  i) $t^2 + 4t - 12$

**5** Factorise these and check your answers.
   a) $x^2 - 4$   b) $y^2 - 25$  c) $z^2 - 1$
   d) $n^2 - 16$  e) $t^2 - 49$  f) $p^2 - 100$

**6** Here is a mixture of types to factorise. Check your answers carefully.
   a) $x^2 + 7x + 6$    b) $x^2 + 6x + 8$   c) $r^2 - 5r + 4$
   d) $x^2 - 8x - 9$    e) $y^2 + 3y - 4$   f) $x^2 + x - 12$
   g) $t^2 - t - 12$    h) $x^2 - 11x + 18$ i) $p^2 + 4p - 12$
   j) $y^2 - 81$        k) $b^2 - b - 20$   l) $a^2 - 11a + 10$

**7** Factorise these completely.
   a) $4x^2 + 4x - 48$  b) $3a^2 - 9a + 6$  c) $3x^2 - 12$
   d) $3x^2 + 6x + 3$   e) $10x^2 - 1000$   f) $5x^2 + 10x - 400$

# 13: Quadratics

# Quadratic equations

Look at the equation

$$x^2 - 2x - 15 = 0$$

This is a **quadratic equation**.

*Why is it called a quadratic equation?*

To solve a quadratic equation, start by factorising the left-hand side.

$$x^2 - 2x - 15 = 0$$
$$(x - 5)(x + 3) = 0$$

This gives you two factors $(x - 5)$ and $(x + 3)$ that multiply to give 0. One of them must be 0.

Either $x - 5 = 0$ and so $x = 5$

or $x + 3 = 0$ and so $x = -3$.

The solution of the equation is $x = 5$ or $-3$.

### Example

Solve $x(x - 4) = -3$

### Solution

First get this equation into the right form.

$$x(x - 4) = -3$$

Expand the brackets $x^2 - 4x = -3$

Add 3 to both sides $x^2 - 4x + 3 = 0$

$$(x - 1)(x - 3) = 0$$

Either $x - 1 = 0$ so $x = 1$

or $x - 3 = 0$ so $x = 3$

The solution is $x = 1$ or $3$

You can get the same result by drawing a graph.

This is the graph of

$y = x^2 - 4x + 3$

You can see that it crosses the $x$ axis at $x = 1$ and at $x = 3$

Sometimes you will meet quadratic equations which cannot be factorised.

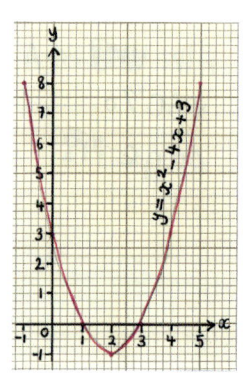

In such cases, you can still solve the equation by drawing the graph.

However, as you will find out from the Investigation on page 149, some quadratic equations have no solution.

## 13: Quadratics

**1** Solve these equations.
  a) $(x-4)(x-1) = 0$
  b) $x(x-4) = 0$
  c) $2(x-4) = 0$
  d) $x(x+4) = 0$
  e) $y(y-3) = 0$
  f) $(x-2)(x-3) = 0$
  g) $(x-5)(x+3) = 0$
  h) $t^2 - 2t = 0$
  i) $y^2 + 4y = 0$

**2** Solve these equations.
  a) $x^2 - 9x + 14 = 0$
  b) $a^2 - 7a + 10 = 0$
  c) $x^2 + 5x - 14 = 0$
  d) $n^2 + 5n + 4 = 0$
  e) $a^2 + a - 6 = 0$
  f) $t^2 - 4t - 12 = 0$
  g) $x^2 + 5x - 50 = 0$
  h) $x^2 - 2x - 63 = 0$
  i) $x^2 - 12x + 11 = 0$

**3** Repeat the steps in question 4 to solve each of these.
  a) $x^2 + x = 20$
  b) $a^2 + 8a = 20$
  c) $r^2 + 4 = 5r$
  d) $y^2 - 3y = 4$
  e) $x^2 + 18 = -11x$
  f) $x^2 = x + 12$
  g) $a^2 + 3a = 16 - 3a$
  h) $x^2 + 3x = 4 + 3x$
  i) $30 + x^2 = 11x$

**4** For each of the following situations
  (i) form an equation in the unknown quantity given
  (ii) solve the equation
  (iii) check your answer.
  a) The length of a rectangular lawn is 3 m greater than its width, $w$ m. Its area is 54 m$^2$.
  b) The width of a box is 10 cm less than its length, $l$ cm. Its height is 8 cm and its volume is 3000 cm$^3$.
  c) A box has a square base of side $x$ cm and height 2 cm. Its outside surface area (including its top and bottom) is 90 cm$^2$.

**5** a) Make out a table of values and draw the graph of $y = x^2 - x$ for values of $x$ from $-4$ to $4$.
  b) Draw the line $y = 6$ on your graph and write down the values of $x$ where the line meets the curve.
  c) Solve the equation $x^2 - x = 6$ as in question 4.
  d) Explain why your answers to b) and c) should be the same.
  e) Use your graph to solve the equation $x^2 - x = 8$.
     Is it possible to solve this by factorising?

### Investigation

Draw the graphs of
  a) $y = x^2 - 6x + 8$
  b) $y = x^2 - 6x + 9$
  c) $y = x^2 - 6x + 10$

Use them to solve the equations
  $x^2 - 6x + 8 = 0$
  $x^2 - 6x + 9 = 0$
  $x^2 - 6x + 10 = 0$

The equation $x^2 - 6x + k = 0$ has no solution. What can you say about $k$?

# 13: Quadratics

# Finishing off

**Now that you have finished this chapter you should be able to**

★ factorise quadratic expressions ★ solve quadratic equations.

Use the questions in the next exercise to check that you understand everything.

## Mixed exercise

**1** Factorise each of these and check your answers.
  a) $x^2 + 13x + 12$
  b) $a^2 - 3a + 2$
  c) $z^2 - 6z + 8$
  d) $n^2 + 11n - 26$
  e) $t^2 - 4t - 12$
  f) $x^2 - 7x - 30$

**2** Factorise each of these and check your answers.
  a) $x^2 + 2x + 1$
  b) $n^2 - 10n + 25$
  c) $r^2 + 8r + 16$
  d) $y^2 - 12y + 36$
  e) $x^2 - 16$
  f) $p^2 - 49$

**3** Solve each of these equations and check your answers.
  a) $2x = 0$
  b) $3(x + 2) = 0$
  c) $x(x - 2) = 0$
  d) $a(a + 15) = 0$
  e) $(t - 5)(t + 6) = 0$
  f) $(a + 7)(a + 11) = 0$
  g) $(x - 3)(x + 3) = 0$
  h) $(2n - 3)(3n - 9) = 0$
  i) $2(y - 5)(3y - 2) = 0$

**4** Solve each of these equations and check your answers.
  a) $x^2 - 5x + 6 = 0$
  b) $b^2 + 7b + 12 = 0$
  c) $x^2 + 18x + 81 = 0$
  d) $x^2 + 3x - 18 = 0$
  e) $d^2 - 4d - 21 = 0$
  f) $x^2 - 10x - 24 = 0$
  g) $x^2 + 2x = 0$
  h) $n^2 - 4 = 0$
  i) $3x^2 - 12 = 0$

**5** Solve each of these equations and check your answers.
  a) $x^2 - x = 56$
  b) $x^2 + x = 90$
  c) $y = y^2 - 12$
  d) $t^2 + 6 = 5t$
  e) $n^2 = 14 - 5n$
  f) $x^2 = 3(6 - x)$
  g) $x^2 = 20 + x$
  h) $x^2 = 5x$
  i) $2x^2 = 20x - 42$

**6** The perimeter of a rectangle is 30 cm and its area is 56 cm².
  a) Draw a rectangle and label one side $x$ cm.
  b) Label the other sides of the rectangle.
  c) Show that $x^2 - 15x + 56 = 0$.
  d) Solve this equation.
  e) State the dimensions of the rectangle.

# 13: Quadratics

**Mixed exercise**

**7** Nikki bought $x - 9$ blouses last week. Each blouse cost £$x$.

The bill came to £36.

a) Show that $x^2 - 9x - 36 = 0$.

b) Solve this equation.

c) How much did each blouse cost?

d) How many blouses did Nikki buy?

e) Why is there only one answer to this problem?

**8** An $n$-sided polygon has $\dfrac{n(n-3)}{2}$ diagonals.

James knows that a particular polygon has 35 diagonals.

a) Show that $n^2 - 3n - 70 = 0$.

b) Solve this equation.

c) State the number of sides of this polygon.

d) Another polygon has 54 diagonals. Calculate the number of sides.

**9** A rectangular garden is $x$ metres by $x + 2$ metres. It has paths and flower beds one metre wide round the edge with a lawn in the middle.

a) Draw a diagram of the garden, showing all the measurements clearly.

b) Write down the dimensions of the lawn in terms of $x$ and find an expression for its area.

c) The lawn requires 120 m$^2$ of turf. Write down an equation for $x$.

d) Solve your equation to find $x$. What are the dimensions of the garden?

---

Design an exhibition space for showing paintings using 16 m of screens. Describe what you intend to exhibit and your reasons for using the screens as you have.

Draw a plan of your space showing any other furniture you might need.

# Fourteen

# Transformations

**Before you start this chapter you should be able to**

★ recognise different types of symmetry

★ recognise congruent shapes

★ decide whether a shape or pattern has rotational symmetry, and if so, its order

★ draw the reflection of a shape in the *x* axis or the *y* axis

★ rotate a shape about its centre or the origin through $\frac{1}{4}$, $\frac{1}{2}$ or $\frac{3}{4}$ turn

★ carry out and describe a translation of a shape.

## Reminder

Shape A has been **reflected** in the mirror line to give shape P.

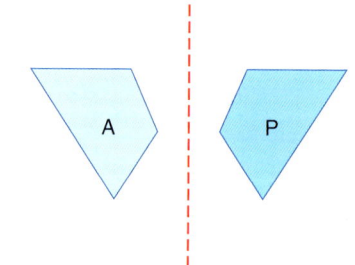

Shape A has been **rotated** through $\frac{1}{4}$ turn clockwise about the point O to give shape Q.

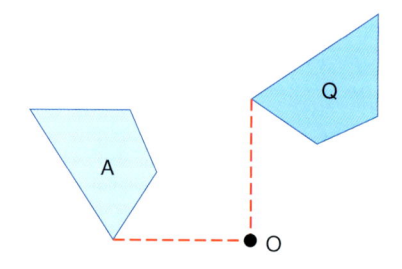

Shape A has been **translated** 3 units to the right and 1 unit down to give shape R.

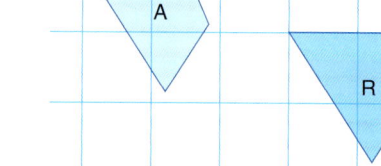

> Shapes A, P, Q and R are all **congruent**. The lengths of lines and the sizes of angles are not changed by these transformations.

Use the questions in the next exercise to check that you still remember these topics.

# 14: Transformations

**Revision exercise**

**1** a) Copy the diagram.

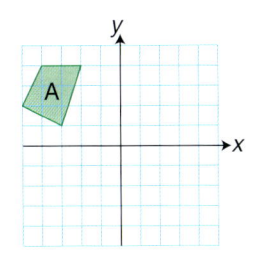

Reflect shape A in the *y* axis. Label the reflection B.
Reflect shape A in the *x* axis. Label the reflection C.

b) Make a second copy of the diagram. Rotate shape A clockwise about the origin through $\frac{1}{4}$ turn, $\frac{1}{2}$ turn and $\frac{3}{4}$ turn. Label the rotations D, E and F respectively.

**2** a) Copy the diagram.

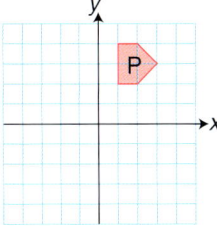

b) Translate shape P 5 squares to the left and 4 squares down. Label the translation Q.

c) Translate shape Q 3 squares to the right and 1 square up. Label this translation R.

d) What translation is needed to move shape R back to the position of shape P?

**3** a) Which of the triangles in this diagram are congruent to A?

b) Which of the triangles are congruent to B?

c) Describe the transformations which map

(i) I → A   (ii) C → D   (iii) E → F

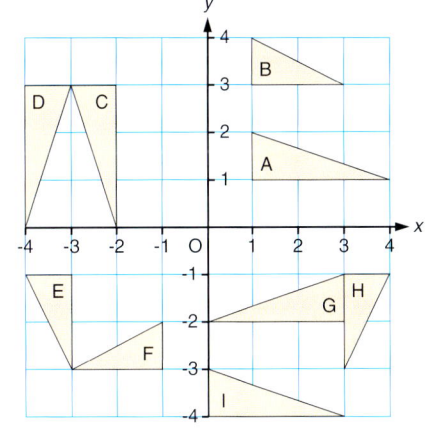

**4** The shape P is formed by joining the points
$(0, 0), (5, 0), (5, 2), (3\frac{1}{2}, 3\frac{1}{2})$

a) Draw P on graph paper. Use the same scales for both *x* and *y*. You will need values between −5 and 5 on both axes.

Now draw the following transformations of P.

b) R, T and V are rotations of P with centre O through 90°, 180° and 270° anticlockwise respectively.

c) W and S are reflections of P in the *x* and *y* axes.

d) Q and U are reflections of P in the lines $y = x$ and $y = -x$.

e) You have now drawn 8 congruent shapes which together make a polygon.

Describe this polygon, and state what symmetry it has.

## 14: Transformations

# Translations using column vectors

Describing a translation by saying how many squares it has moved in each direction is rather long-winded.

A quicker way of describing a translation is to use a **column vector**.

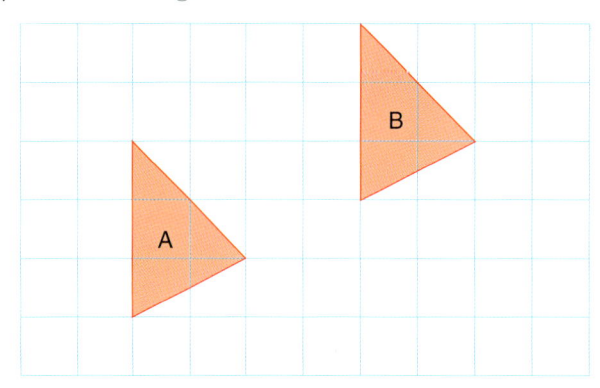

Here is the column vector that describes the translation from A to B:

The top number is the number of units moved to the right

$\begin{pmatrix} 4 \\ 2 \end{pmatrix}$

The bottom number is the number of units moved up

- If the translation is to the left instead of the right, the top number is negative.
- If the translation is down instead of up, the bottom number is negative.

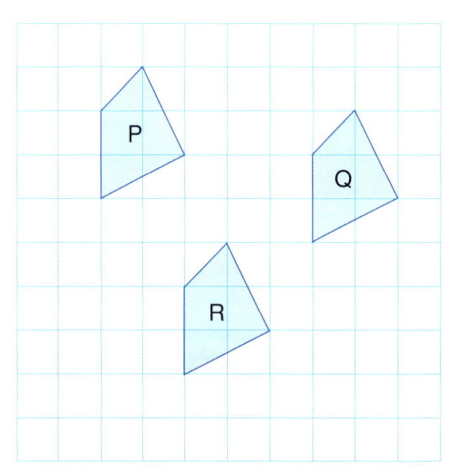

The translation from P to Q is $\begin{pmatrix} 5 \\ -1 \end{pmatrix}$

*What is R to P?*

*What is Q to R?*

# 14: Transformations

**1** Write down the column vectors for the following translations.

a) A to B   b) D to E   c) B to C
d) E to B   e) C to A   f) B to D
g) A to D   h) E to C

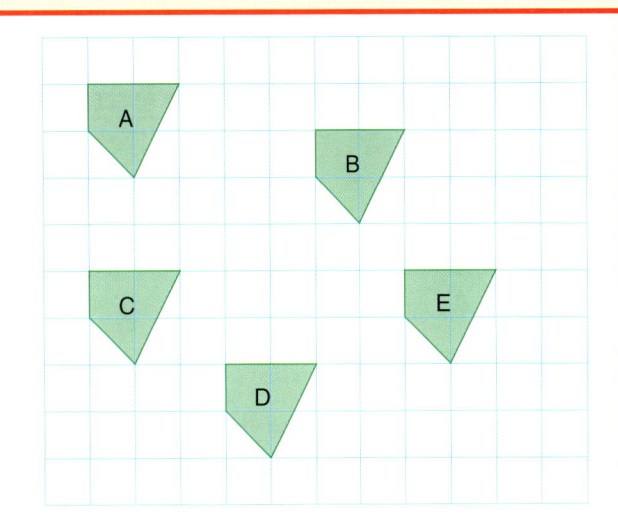

**2** Copy the diagram.

a) Translate shape A using the vector $\begin{pmatrix} -5 \\ 3 \end{pmatrix}$. Label the translated shape B.

b) What vector translates shape B back to shape A?

c) Translate shape B using the vector $\begin{pmatrix} -2 \\ -4 \end{pmatrix}$. Label the translated shape C.

d) What vector translates shape C back to shape B?

e) What vector translates shape A to shape C?

f) What vector translates shape C to shape A?

**3** Look back at your answers to question 2.

a) What is the rule connecting a vector with its reverse (e.g. A→B and B→A)?

b) Compare the vector A→C with the vectors A→B and B→C. What do you notice?

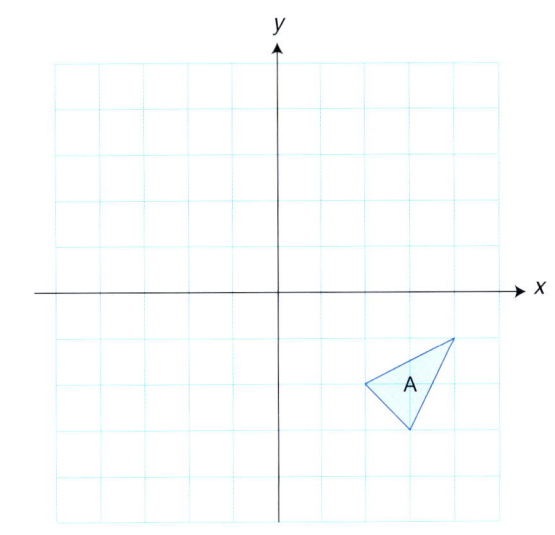

---

For chess players only.
A knight is on one of the centre squares of an empty chess board. There are 8 squares that it can move to; one of these needs the translation $\begin{pmatrix} 2 \\ -1 \end{pmatrix}$.

Describe the other moves.
Now the knight is placed in one of the corners. Which square of the board takes the greatest number of moves for the knight to reach it?

155

# 14: Transformations

# Reflection

## Reflection in a vertical or horizontal line

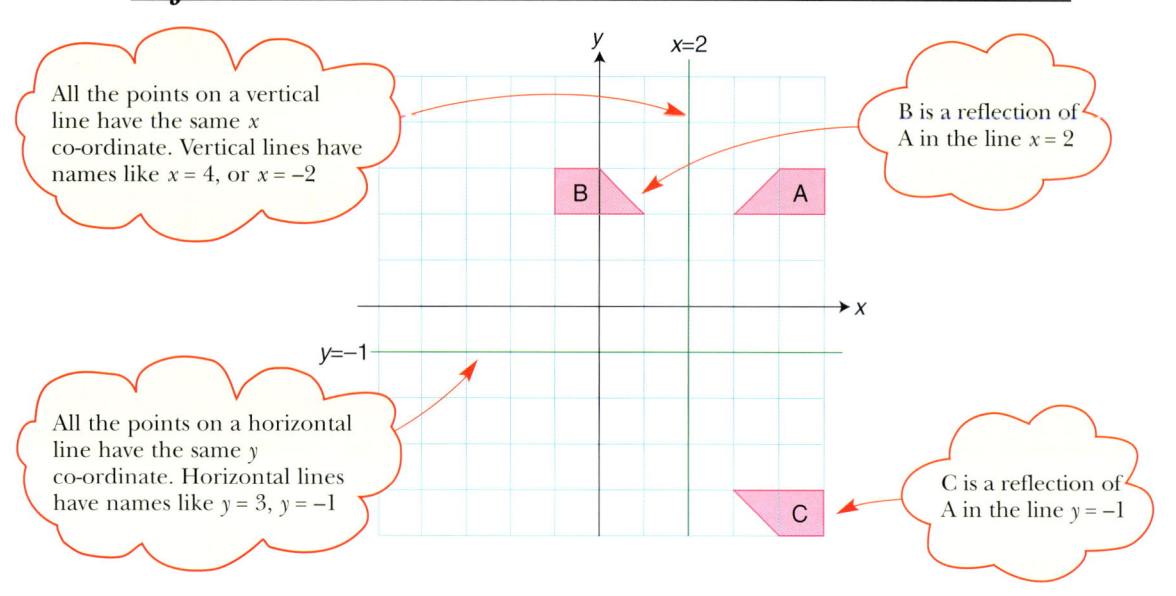

All the points on a vertical line have the same $x$ co-ordinate. Vertical lines have names like $x = 4$, or $x = -2$

B is a reflection of A in the line $x = 2$

All the points on a horizontal line have the same $y$ co-ordinate. Horizontal lines have names like $y = 3$, $y = -1$

C is a reflection of A in the line $y = -1$

## Reflection in a diagonal line

When you draw a reflection in a diagonal line, it is best to reflect its corners one at a time. The diagram shows you how to do this.

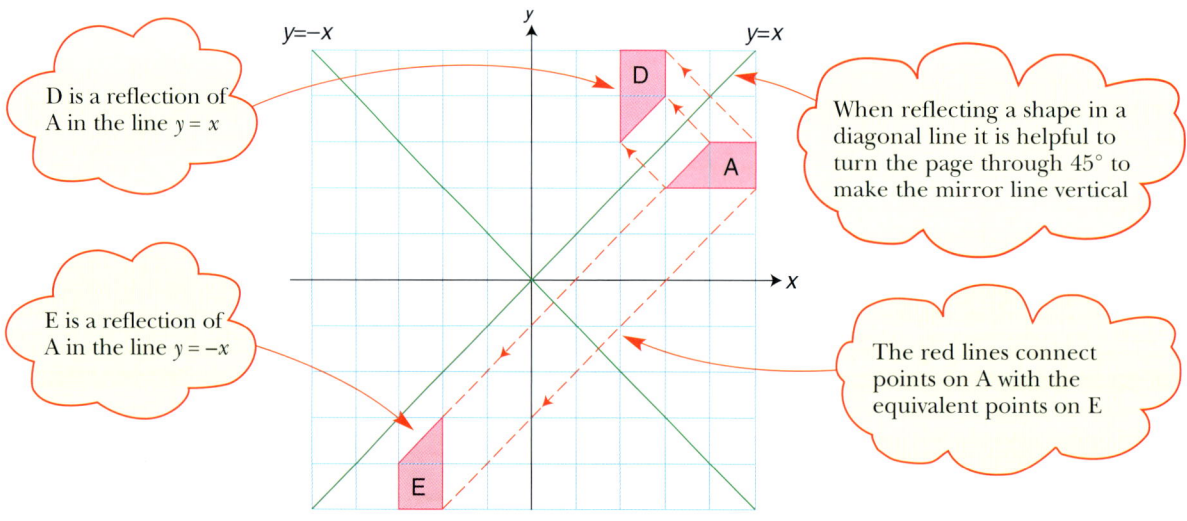

D is a reflection of A in the line $y = x$

When reflecting a shape in a diagonal line it is helpful to turn the page through 45° to make the mirror line vertical

E is a reflection of A in the line $y = -x$

The red lines connect points on A with the equivalent points on E

? Why are the names of the two diagonal lines $y = x$ and $y = -x$?
(Think about the co-ordinates of points on the lines.)

156

# 14: Transformations

**1** Copy these diagrams onto squared paper and draw the reflection of each shape in the mirror line shown.

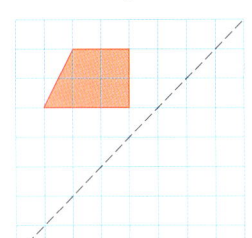

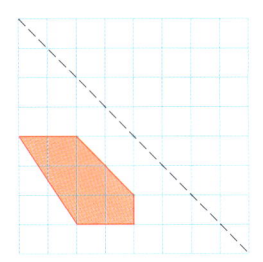

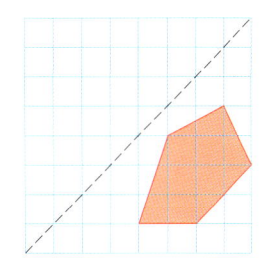

**2** Make 4 copies of this diagram on squared paper. Reflect the triangle in each of these mirror lines.

   a) $y = 1$    b) $x = 2$
   c) $y = -2$   d) $x = -1$

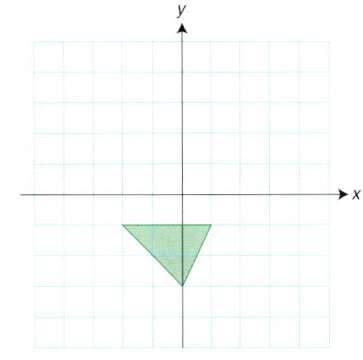

**3** a) Write down the co-ordinates of each corner of triangle A.

   b) Copy the diagram and reflect triangle A in the line $y = x$. Label this triangle B.

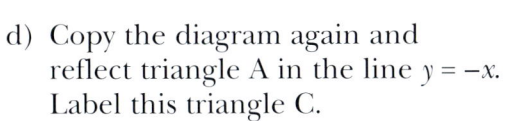

   c) Write down the co-ordinates of each corner of triangle B. What is the rule connecting the co-ordinates of A with the co-ordinates of B?

   d) Copy the diagram again and reflect triangle A in the line $y = -x$. Label this triangle C.

   e) Write down the co-ordinates of each corner of triangle C. What is the rule connecting the co-ordinates of A with the co-ordinates of C?

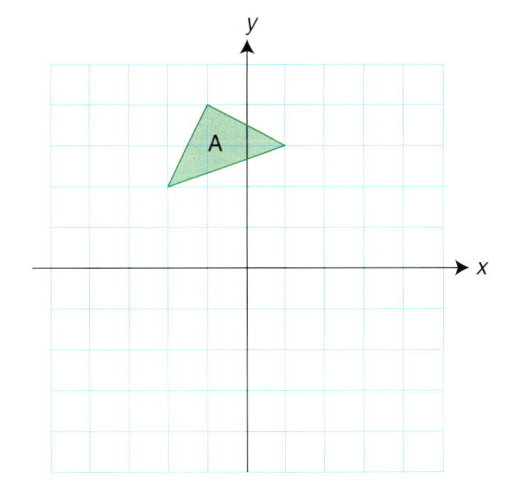

---

Which letters of the alphabet have mirror symmetry
a) about the $x$ axis?
b) about the $y$ axis?
State a word which has each type of symmetry.
Find out the meaning of the word palindromic.

## 14: Transformations

# Rotation

You already know how to rotate a shape about the origin. The diagram shows how to rotate a shape about any other point.

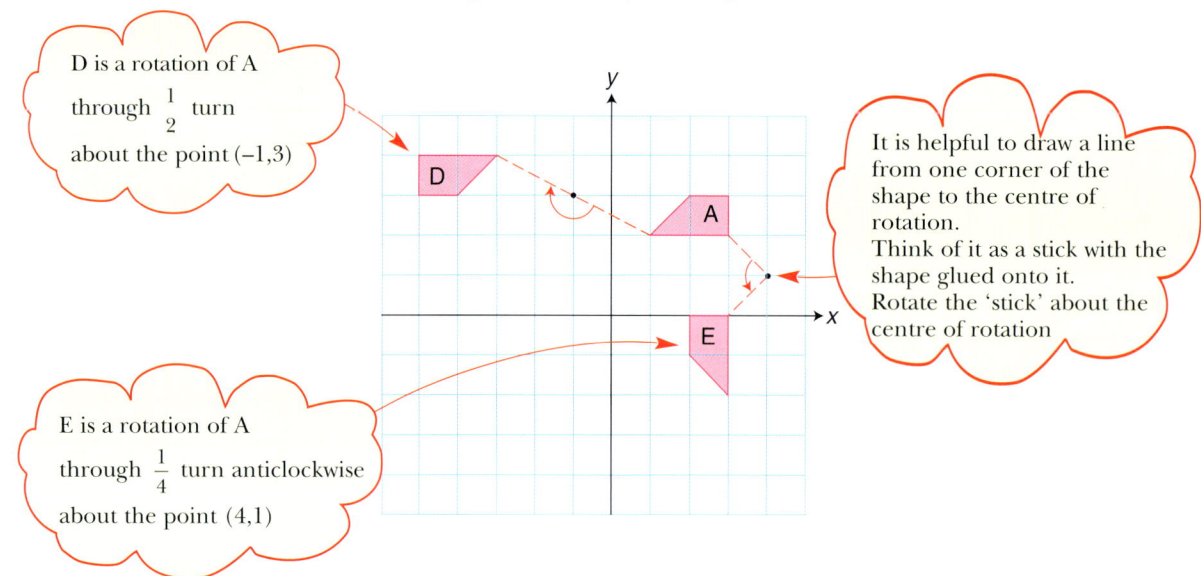

D is a rotation of A through $\frac{1}{2}$ turn about the point (−1,3)

It is helpful to draw a line from one corner of the shape to the centre of rotation.
Think of it as a stick with the shape glued onto it.
Rotate the 'stick' about the centre of rotation

E is a rotation of A through $\frac{1}{4}$ turn anticlockwise about the point (4,1)

## *Recognising reflections, rotations and translations*

Sometimes it is easy to recognise a reflection, rotation or translation and to spot the mirror line or centre of rotation. In cases where you are not sure, try joining up pairs of corresponding points.

For a translation, the lines will be parallel and the shapes will be the same way round.

For a reflection, the lines will be parallel and the shape will be 'flipped over'. The mirror line goes down the middle.

For a rotation through $\frac{1}{2}$ turn, the lines all meet at a point.

? What is special about this point?

If the lines are not parallel and do not meet at a point, you need to check for $\frac{1}{4}$ or $\frac{3}{4}$ turn. (In some cases it is hard to find the centre of rotation, but not in the examples in this book.)

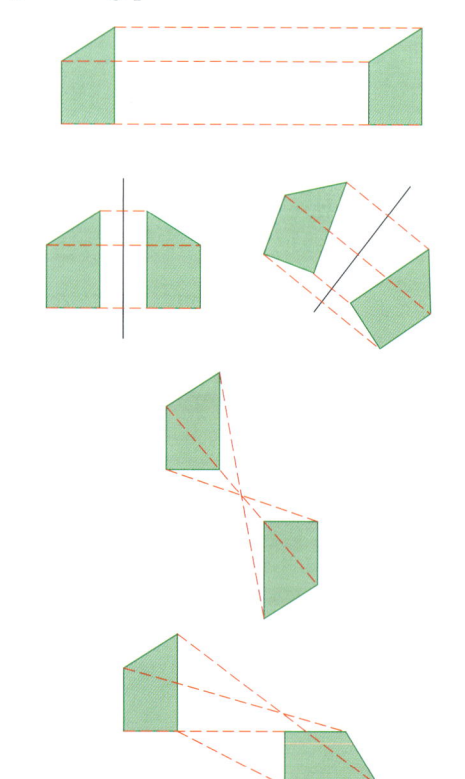

# 14: Transformations

**1** Copy this diagram.

For each of these rotations, draw the rotation and label it clearly.

a) $\frac{1}{4}$ turn clockwise about (−4, 1)

b) $\frac{3}{4}$ turn clockwise about (1, 2)

c) $\frac{1}{2}$ turn about (1, 0)

d) $\frac{1}{4}$ turn anticlockwise about (−1, 1)

e) $\frac{1}{2}$ turn about (1, −1)

**2** For each of the following transformations, state whether it is a reflection, a rotation or a translation. For a reflection, give the equation of the mirror line, for a rotation, give the co-ordinates of the centre of rotation, and for a translation, give the translation vector.

a) A to B   b) B to C   c) C to D

d) D to E   e) E to F   f) F to G

g) G to H   h) H to I   i) I to A

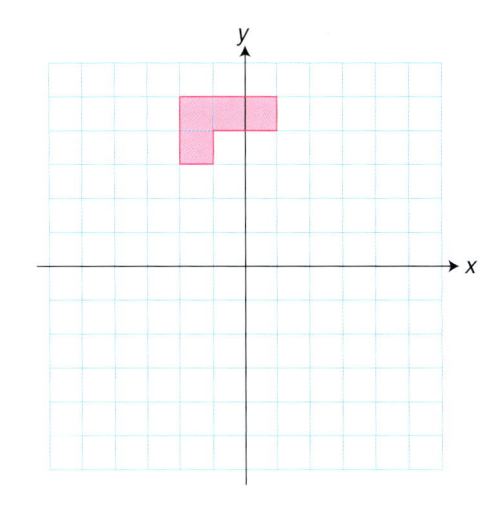

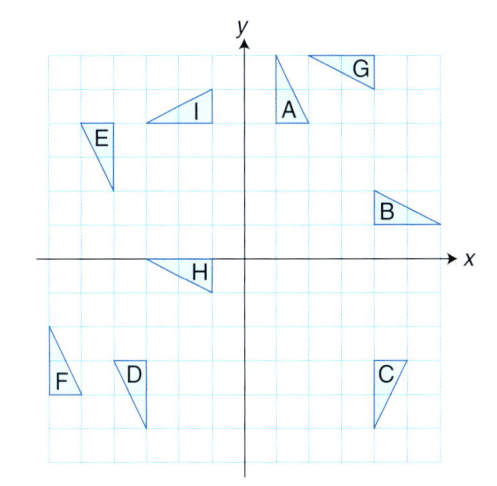

How many single transformations can you find which map A→B?

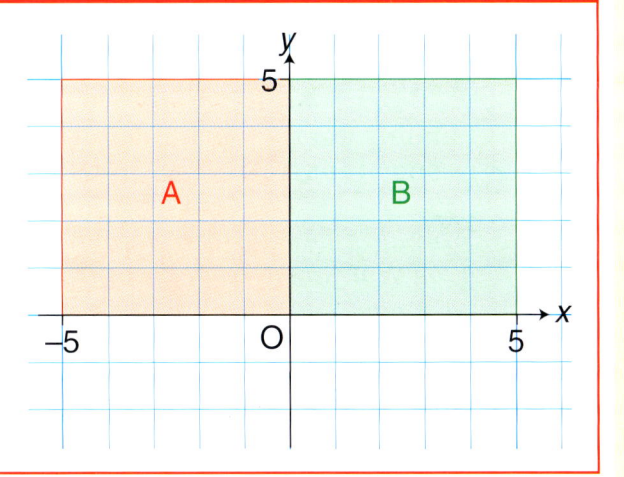

159

### 14: Transformations

# Combining transformations

This diagram shows part of a repeating pattern. It is made out of transformations of triangle A.

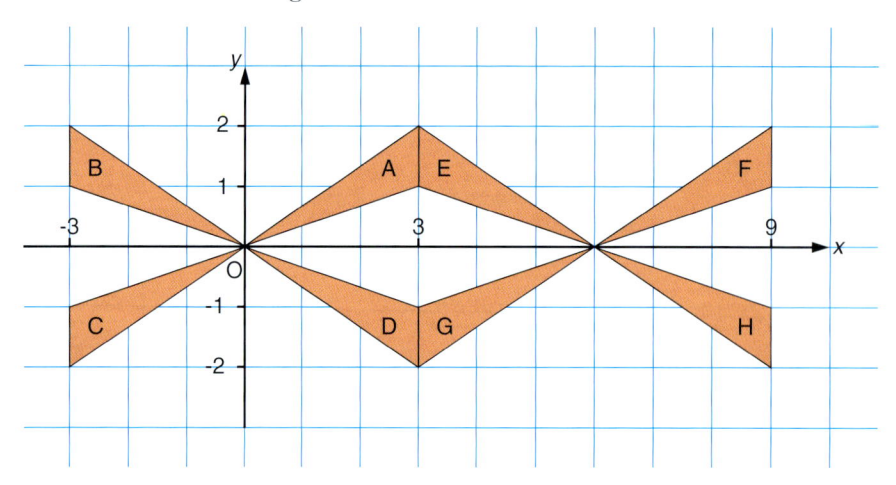

> *You can map A onto each of B, C, D, E, F and G by a single transformation. In each case it is either a reflection or a rotation or a translation. How would you describe these six transformations?*

When you come to H, you find that there is no single transformation. You need two. One way of doing this is

> First rotate A centre (0, 0) through 180° then do a reflection in the vertical line $x = 3$.

> *How many more ways can you find to map A → H in two steps? Remember to be careful to say which of the two comes first and which second.*

Here is a real-life example of combining transformations.

The diagram shows a 400 m running track.

An athlete starts at point A, (0, 32) and runs round the track in a clockwise direction.

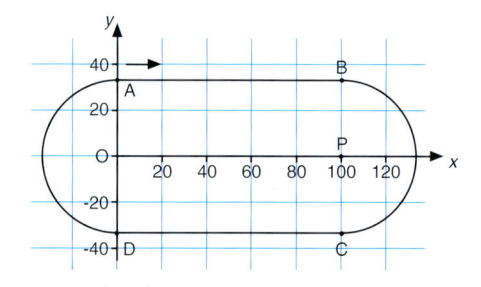

Running from A to B involves a translation $\begin{pmatrix} 100 \\ 0 \end{pmatrix}$ and from B to C a rotation centre P (100, 0) through 180°.

> *What transformations represent the athlete's run from C back to A? Which of the four transformations cancel each other out? Can the athlete have a reflection?*

# 14: Transformations

Questions 1–4 of this exercise refer to the triangles in this diagram.

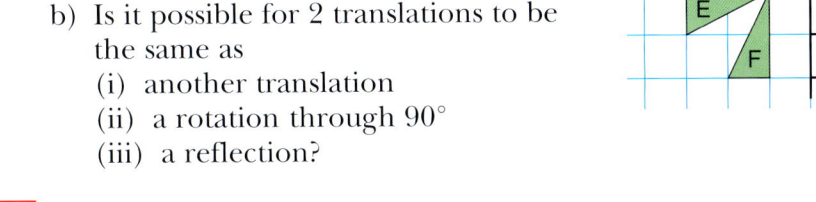

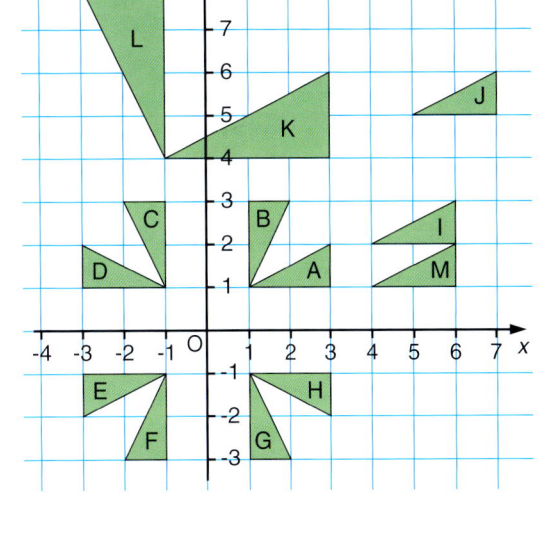

**1** Describe the transformations which map

a) (i) A → D  (ii) D → E  (iii) A → E

b) (i) A → D  (ii) D → M  (iii) A → M

c) Is it possible for 2 reflections to be the same as
   (i) a translation
   (ii) a rotation
   (iii) another reflection?

**2** Describe the transformations which map

a) (i) A → I  (ii) I → J  (iii) A → J

b) Is it possible for 2 translations to be the same as
   (i) another translation
   (ii) a rotation through 90°
   (iii) a reflection?

**3** Describe the transformations which map

a) (i) A → K  (ii) K → L

b) (i) A → C  (ii) C → L

c) Is there a single transformation which maps A → L?

**4** Give examples using the triangles in the diagram, to show that

a) two rotations are equivalent to another rotation.

b) the same reflection carried out twice cancels itself out.

---

The diagram shows a settee in one corner of a room (position A) which has to be turned round and moved to the opposite corner (position B). The settee has legs at its four corners. It is too heavy to lift or slide and so has to be rotated about one of its legs.

Copy the diagram and shade, in different colours, the moves needed to get the settee to its new position. Mark in the centre of each rotation.

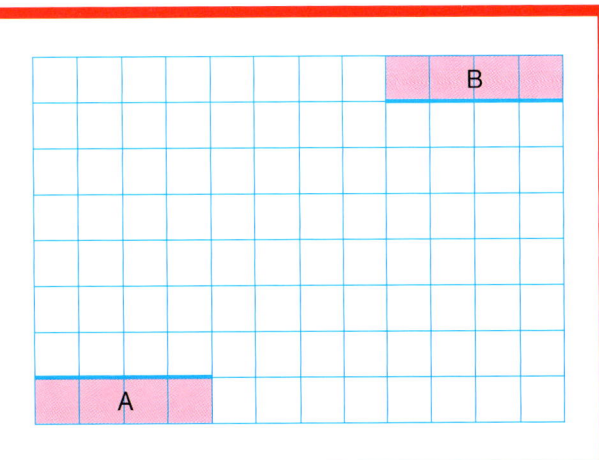

# 14: Transformations

## Finishing off

**Now that you have finished this chapter you should be able to**

★ recognise and draw a reflection of a simple shape in horizontal, vertical or diagonal lines

★ rotate a simple shape about any point through $\frac{1}{4}$ turn, $\frac{1}{2}$ turn or $\frac{3}{4}$ turn

★ find the angle of rotation and centre of rotation given a shape and its rotated image

★ understand and use a column vector to describe a translation

★ investigate situations combining two transformations.

Use the questions in the next exercise to check that you understand everything.

## Mixed exercise

**1** For each of these, copy the diagram and draw the transformation:

a) reflection in the line $x = -1$

b) reflection in the line $y = 2$

c) reflection in the line $y = -x$

d) rotation through $\frac{1}{4}$ turn anticlockwise about the point (5, 2)

e) rotation through $\frac{1}{2}$ turn about the point (−1, 1)

f) translation using the vector $\begin{pmatrix} -4 \\ -5 \end{pmatrix}$.

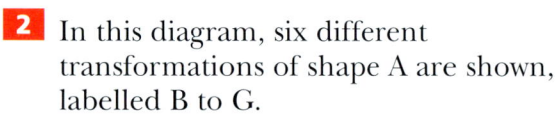

**2** In this diagram, six different transformations of shape A are shown, labelled B to G.

Describe each of the transformations fully.

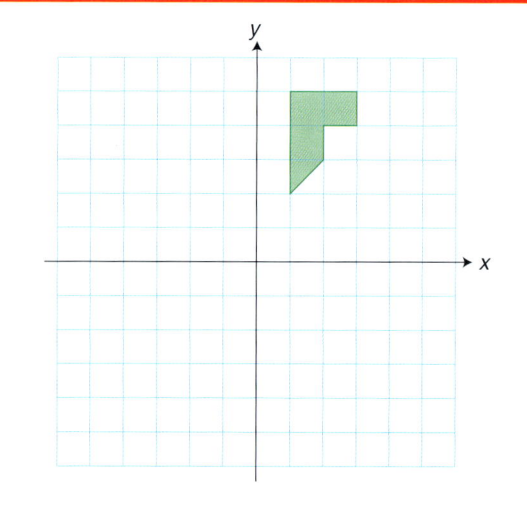

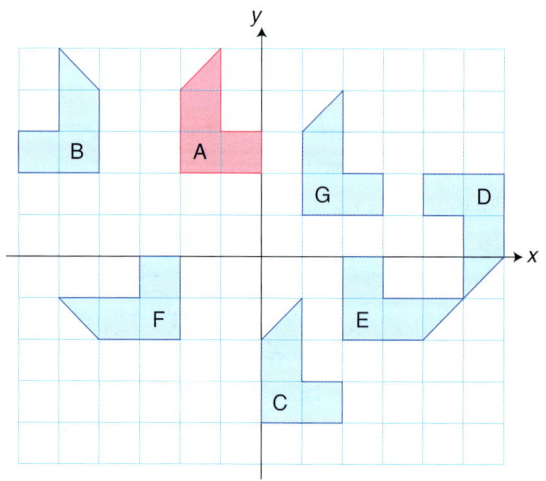

# 14: Transformations

**Mixed exercise**

**3** A triangle A has co-ordinates (0, 0), (1, 0), (3, 3).

a) Draw triangle A, using the same scale for both $x$ and $y$ axes. Take values between −3 and +3 for both $x$ and $y$.

b) Triangles B, C and D are formed by rotating A anticlockwise about O through 90°, 180° and 270°, respectively. Draw, B, C and D.

c) Triangles E, F, G and H are formed by reflecting A, B, C and D in the $x$ axis. Draw E, F, G and H.

d) Describe single transformations which map

(i) A → G   (ii) A → H.

**4** Answer this question on graph paper. Use the same scale for the $x$ axis (values −4 to 12) and the $y$ axis (values 0 to 12).

A triangle L has co-ordinates (4, 3), (6, 3), (6, 4).
Another triangle P has co-ordinates (0, 6), (4, 6), (4, 8).

a) Find the centre of the enlargement which maps L onto P, and the scale factor.

b) Triangle P is given a translation $\begin{pmatrix} 4 \\ 0 \end{pmatrix}$ to form triangle Q, and a translation $\begin{pmatrix} 8 \\ 0 \end{pmatrix}$ to form R. Find the centres of the enlargements which map L onto Q and R.

c) Triangle L is now enlarged from the same three centres but with scale factor 3. Do the enlarged triangles touch each other, like P, Q and R do?

**5** The diagram shows an equilateral triangle A. Copy it onto graph paper.

a) On the same diagram as A, draw triangle B which is formed by rotating A through 60° anticlockwise about O.

b) Add in triangles C, D, E and F formed by rotating A through 120°, 180°, 240° and 300° about O.

c) Describe the complete shape you have now drawn.

d) Is it possible to draw the same shape starting with A and doing reflections in the $x$ axis and reflections in the $y$ axis?

e) Describe how you can draw the same shape starting with A and doing translations and reflections in the $x$ axis.

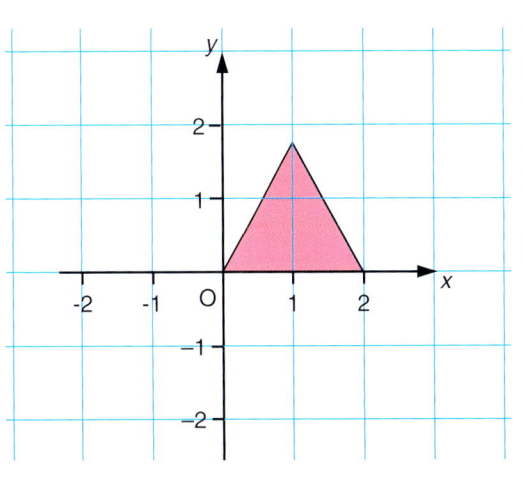

163

# Fifteen

# Fractions in algebra

> **Before you start this chapter you should be able to**
> 
> ★ find equivalent fractions and simplest forms
> ★ use indices
> ★ multiply and divide powers of a number
> ★ solve equations containing fractions.

## Reminder

You can multiply (or divide) the top and the bottom of a fraction by the same number.

*4 is 2 × 2*
*10 is 5 × 2*

$$\frac{2}{5} = \frac{4}{10} = \frac{8}{20} = \frac{24}{60}$$

$$\frac{2x}{5x} = \frac{2}{5} = \frac{8x}{20x}$$

$$\frac{x+1}{3} + 2 = 8$$
$$\frac{x+1}{3} = 6$$
$$x+1 = 18 \quad \text{so } x = 17$$
$$\text{Check } \frac{17+1}{3} + 2 = 6 + 2 = 8 \checkmark$$

Use the questions in the next exercise to check that you still remember these topics.

## Revision exercise

*Find the missing numbers in questions 1 to 4.*

**1**
a) $\frac{1}{2} = \frac{?}{16}$
b) $\frac{1}{3} = \frac{?}{27}$
c) $\frac{2}{3} = \frac{?}{18}$
d) $\frac{3}{4} = \frac{?}{12}$

e) $\frac{6}{10} = \frac{?}{5}$
f) $\frac{36}{96} = \frac{?}{8}$
g) $\frac{28}{49} = \frac{4}{?}$
h) $\frac{15}{39} = \frac{?}{13}$

**2**
a) $\frac{1}{2} = \frac{8}{?}$
b) $\frac{1}{4} = \frac{3}{?}$
c) $\frac{2}{3} = \frac{10}{?}$
d) $\frac{3}{5} = \frac{9}{?}$

e) $\frac{21}{49} = \frac{3}{?}$
f) $\frac{3}{18} = \frac{1}{?}$
g) $\frac{21}{28} = \frac{3}{?}$
h) $\frac{36}{42} = \frac{6}{?}$

## 15: Fractions in algebra

**Revision exercise**

**3** a) $\dfrac{x}{2} = \dfrac{8x}{?}$   b) $\dfrac{x}{4} = \dfrac{3x}{?}$   c) $\dfrac{2x}{3} = \dfrac{10x}{?}$   d) $\dfrac{3y}{5} = \dfrac{9y}{?}$

e) $\dfrac{21y}{49} = \dfrac{3y}{?}$   f) $\dfrac{3y}{18} = \dfrac{y}{?}$   g) $\dfrac{21z}{28} = \dfrac{3z}{?}$   h) $\dfrac{36z}{42} = \dfrac{6z}{?}$

**4** a) $\dfrac{1}{2x} = \dfrac{?}{16x}$   b) $\dfrac{1}{3x} = \dfrac{?}{27x}$   c) $\dfrac{2}{3x} = \dfrac{?}{18x}$   d) $\dfrac{3}{4y} = \dfrac{?}{12y}$

e) $\dfrac{6}{10y} = \dfrac{?}{5y}$   f) $\dfrac{36}{96z} = \dfrac{?}{8z}$   g) $\dfrac{28}{49z} = \dfrac{?}{7z}$   h) $\dfrac{15}{39z} = \dfrac{?}{13z}$

**5** a) Write True or False in each case.
   $\dfrac{3x}{12}$ is equivalent to ...

   (i) $\dfrac{9x}{36}$   (ii) $\dfrac{1}{4}$   (iii) $\dfrac{x}{4}$   (iv) $\dfrac{3}{12}$   (v) $\dfrac{3}{12x}$   (vi) $\dfrac{4x}{16}$

b) Find the missing term in each of these.

   (i) $\dfrac{x}{4} = \dfrac{?}{16}$   (ii) $\dfrac{4x}{24} = \dfrac{?}{6}$   (iii) $\dfrac{x}{3} = \dfrac{?}{21}$   (iv) $\dfrac{5x}{12} = \dfrac{?}{84}$

   (v) $\dfrac{3x}{11} = \dfrac{?}{121}$   (vi) $\dfrac{6x}{10} = \dfrac{?}{5}$   (vii) $\dfrac{6x}{36} = \dfrac{?}{6}$   (viii) $\dfrac{36x}{48} = \dfrac{?}{4}$

*Solve the equations in questions 6 to 9.*

**6** a) $\dfrac{a}{4} = 7$   b) $\dfrac{b}{3} = 4$   c) $\dfrac{c}{4} = 4$

d) $\dfrac{2d}{3} = 6$   e) $\dfrac{4e}{5} = 20$   f) $\dfrac{3f}{7} = 21$

**7** a) $\dfrac{a}{4} + 3 = 9$   b) $\dfrac{b}{3} + 6 = 8$   c) $\dfrac{c}{5} + 4 = 7$

d) $\dfrac{d}{3} - 1 = 5$   e) $\dfrac{e}{5} - 6 = 3$   f) $\dfrac{f}{2} + 3 = 1$

**8** a) $\dfrac{x+1}{3} + 3 = 5$   b) $\dfrac{x+2}{7} + 1 = 4$   c) $\dfrac{x-3}{5} + 2 = 3$

d) $\dfrac{x-5}{5} - 1 = 3$   e) $\dfrac{2x+1}{5} - 2 = 3$   f) $\dfrac{3x+2}{2} + 1 = 11$

**9** a) $\dfrac{5}{x} = \dfrac{20}{16}$   b) $\dfrac{3}{x} = \dfrac{18}{12}$   c) $\dfrac{5}{x} = \dfrac{20}{4}$

d) $\dfrac{2(x+1)}{3} = 12$   e) $\dfrac{5(x-2)}{4} = \dfrac{15}{2}$   f) $\dfrac{3(x-3)}{8} = \dfrac{9}{4}$

165

## 15: Fractions in algebra

# Indices

### Do you remember?

The power or **index** tells you how many numbers you have.

*There are three 8s. The index is 3*

$8 \times 8 \times 8 = 8^3$

To multiply numbers you *add* the indices.

*4 + 3 = 7*

$8^4 \times 8^3 = (8 \times 8 \times 8 \times 8) \times (8 \times 8 \times 8) = 8^7$

To divide numbers you *subtract* the indices.

*5 − 3 = 2*

$8^5 \div 8^3 = \dfrac{8 \times 8 \times 8 \times 8 \times 8}{8 \times 8 \times 8} = 8^2$

$6c^3 \times 4c^2 = 6 \times c \times c \times c \times 4 \times c \times c = 24c^5$

$\dfrac{6c^5}{4c^3} = \dfrac{6 \times c \times c \times \cancel{c} \times \cancel{c} \times \cancel{c}}{4 \times \cancel{c} \times \cancel{c} \times \cancel{c}} = \dfrac{3c^2}{2}$

**?** What is $y^{49} \times y^{51}$?

**?** What is $z^{26} \div z^{17}$?

### Class discussion

$a$ and $n$ are whole numbers.

What values of $a$ and $n$ make $a^n$ close to 123?

## 15: Fractions in algebra

**1** Write down each of these using indices.
   a) (i) $5 \times 5 \times 5 \times 5$   (ii) $7 \times 7 \times 7 \times 7 \times 7 \times 7$   (iii) $3 \times 3 \times 3 \times 3 \times 3$
   b) (i) $a \times a \times a \times a$   (ii) $b \times b \times b \times b \times b \times b$   (iii) $c \times c \times c \times c \times c$

**2** Write each of these out in full.
   a) $9^3$    b) $7^4$    c) $5^8$    d) $4^2$    e) $3^5$    f) $6^7$
   g) $a^3$    h) $b^4$    i) $c^8$    j) $3d^2$    k) $7x^5$    l) $4y^7$

**3** Write these as briefly as possible using indices.
   a) $3^2 \times 3^5$    b) $5^4 \times 5^3$    c) $4^{12} \div 4^{10}$    d) $3^2 \times 3^3 \div 3^4$
   e) $x^2 \times x^5$    f) $y^4 \times y^4$    g) $a^{11} \div a^7$    h) $b^4 \times b^2 \div b^5$

**4** Write these as briefly as possible using indices.
   a) $2x \times 4$    b) $3y \times 7$    c) $5z \times 3$    d) $9x^2 \times 2$
   e) $3x \times 4x$    f) $2y \times 6y^2$    g) $5z \times 6z^3$    h) $8x^2 \times 2x$

**5** Simplify these.
   a) $a^5 \div a^3$    b) $b^4 \div b^3$    c) $c^7 \div c^3$    d) $d^4 \div d^2$
   e) $x^9 \div x^4$    f) $x^7 \div x^3$    g) $x^{10} \div x^5$    h) $y^8 \div y^3$

**6** Simplify these.
   a) $12a^5 \div 4a^3$    b) $36b^4 \div 9b^3$    c) $21c^7 \div 7c^3$    d) $15d^4 \div 5d^2$
   e) $4x^9 \div 16x^4$    f) $5x^7 \div 25x^3$    g) $6x^{10} \div 18x^5$    h) $5y^8 \div 20y^3$

**7** Simplify these.
   a) $\dfrac{x^4 \times x^5}{x^8}$    b) $\dfrac{x^3 \times x^6}{x^7}$    c) $\dfrac{a^4 \times a^6}{a^5}$    d) $\dfrac{a^7 \times a^5}{a^{10}}$
   e) $\dfrac{3a \times 4b^2}{2a^2 b}$    f) $\dfrac{6cd^3}{15d^2}$    g) $\dfrac{21a^2 b^2 c^2}{15ac^3}$    h) $\dfrac{8r^2 s^2}{2rs^2}$

---

Make a list of powers of 2 and use it to do the following problems.
Check your answers using a calculator

   a) $\dfrac{256}{64}$    b) $\left(\dfrac{128}{32}\right)^3$    c) $\dfrac{256 \times 32}{128}$    d) $\sqrt{\dfrac{2048}{512}}$

Make up some similar problems and exchange them with a friend.

# 15: Fractions in algebra

## Rational functions

Steve and Ray are doing their algebra homework.

Here is one of the questions.

Simplify $\dfrac{x^2 + 5x - 6}{x + 6}$

*An algebra fraction like this is called a **rational function***

This is what Steve wrote.

Ray did it like this.

$\dfrac{x^2 + 5x - 6}{x + 6}$

$= \dfrac{x^2 + 5x - 6}{x + 6}$

$= x^2 + 5$ ✗

$\dfrac{x^2 + 5x - 6}{x + 6}$

$= \dfrac{(x-1)(x+6)}{x+6}$

$= x - 1$ ✓

You must *never* cancel bits of an expression as Steve did. If you had $\dfrac{16}{68}$ you wouldn't cancel the 6s. You would look for a common factor.

You would write

$\dfrac{16}{68} = \dfrac{4 \times 4}{4 \times 17} = \dfrac{4}{17}$

*You can cancel these 4s.*

*well that's certainly a lot simpler*

**?** Why can't you cancel $\dfrac{x-1}{x+1}$ ?

Try several even-number values for x and check what happens.

**?** Can you simplify $\dfrac{(x-1)(x+1)(x+2)}{(x+2)(x-3)(x+1)}$ ?

## 15: Fractions in algebra

**Exercise**

**1** Simplify these.

a) $\dfrac{x(x+1)}{(x+1)}$    b) $\dfrac{(x+1)x}{(x+1)}$    c) $\dfrac{x^2(x-2)}{(x-2)}$

d) $\dfrac{(x-1)x^3}{x(x-1)}$    e) $\dfrac{(x+4)x^4}{x^3(x+4)}$    f) $\dfrac{(2x-1)x^3}{x^4(2x-1)}$

**2** a) Factorise (i) $3x^2 + x$    (ii) $4x^3 + x^2$    (iii) $5x^2 + x^3$

b) Use part a) to simplify (i) $\dfrac{3x^2+x}{x}$    (ii) $\dfrac{4x^3+x^2}{x^2}$    (iii) $\dfrac{5x^2+x^3}{x^2}$

**3** Simplify these.

a) $\dfrac{(x+1)(x-5)}{(x+1)}$    b) $\dfrac{(x+1)(x-2)}{(x+1)}$    c) $\dfrac{(x+3)(x-2)}{(x-2)}$

d) $\dfrac{(x-1)(x+1)}{(x-1)}$    e) $\dfrac{(x+4)(x-2)}{(x+4)}$    f) $\dfrac{(2x-1)(x-2)}{(2x-1)}$

**4** Simplify these.

a) $\dfrac{(x+1)(x-5)}{(x+1)(x+5)}$    b) $\dfrac{(x+1)(x-2)}{(x+2)(x+1)}$    c) $\dfrac{(x+3)(x-2)}{(x-2)(x-3)}$

**5** a) Factorise (i) $x^2 + 3x$ (ii) $x^2 - 1$ (iii) $x^2 - 9x$ (iv) $3x^2 + 2x$ (v) $x^3 + x^2$

b) Use part a) to simplify (i) $\dfrac{x^2+3x}{x+3}$    (ii) $\dfrac{x^2-1}{x+1}$

(iii) $\dfrac{x^2-9x}{x-9}$    (iv) $\dfrac{x^2}{3x^2+2x}$    (v) $\dfrac{x^3}{x^3+x^2}$

**6** a) Factorise (i) $x^2 + 7x + 10$    (ii) $x^2 - 5x + 6$    (iii) $x^2 + 2x - 3$

b) Use part a) to simplify

(i) $\dfrac{x^2+7x+10}{x+2}$    (ii) $\dfrac{x^2-5x+6}{x-3}$    (iii) $\dfrac{x^2+2x-3}{x-1}$

---

Try simplifying these. You need to factorise the top and the bottom of the fraction.

a) $\dfrac{x^2+7x+10}{x^2-25}$    b) $\dfrac{x^2-x+2}{x^2-4x+4}$    c) $\dfrac{x^2-2x-3}{x^2-1}$

d) $\dfrac{x^2+x-2}{x^2+3x+2}$    e) $\dfrac{x^2-8x+4}{x^2+3x-10}$    f) $\dfrac{x^2+4x+3}{x^2-x-2}$

## 15: Fractions in algebra

# Finishing off

**Now that you have finished this chapter you should be able to**

★ find equivalent fractions in algebra
★ simplify expressions by using indices
★ use indices to multiply and divide
★ cancel a common factor in rational functions.

Use the questions in the next exercise to check that you understand everything.

**Mixed exercise**

*In questions 1 to 3, write the expressions as briefly as possible using indices*

**1** a) $5x^2 \times 4x^3$  b) $3y^3 \times 6y^2$  c) $4z \times 3z^7$  d) $9x^2 \times 4x^3$
   e) $3x \times 4x \times 5x$  f) $3y \times 5y^2 \times y$  g) $5z \times 9z \times z^2$  h) $5x^8 \times 2x \times 3x^2$

**2** a) $y^6 \div y^3$  b) $z^{12} \div z^{10}$  c) $z^7 \div z^2$  d) $z^5 \div z^2$
   e) $12y^6 \div 8y^3$  f) $36z^{12} \div 30z^{10}$  g) $12z^7 \div 18z^2$  h) $22z^5 \div 33z^2$

**3** a) $\dfrac{4y^4 \times 2y^2}{y^3}$  b) $\dfrac{3y^4 \times 5y^7}{y^5}$  c) $\dfrac{4z^6 \times 5z^7}{z^8}$  d) $\dfrac{3z^5 \times 3z^{11}}{z^9}$
   e) $\dfrac{4x^4 \times 5x^5}{8x^8}$  f) $\dfrac{10x^3 \times 3x^7}{15x^9}$  g) $\dfrac{6y^3 \times 5y^4}{10y^5}$  h) $\dfrac{8y^2 \times 3y^6}{12y^5}$

**4** Simplify these.

a) $\dfrac{3x^2}{x^5}$  b) $\dfrac{4x^3y^2}{2xy^4}$  c) $\dfrac{9x^4y^3z^9}{3x^4y^2z^4}$  d) $\dfrac{15a^3b^2c}{5ab^2c^3}$

e) $\dfrac{24ab^7c^4}{18a^6b^2c^3}$  f) $\dfrac{32a^2b^2c^2}{56a^3bc^7}$  g) $\dfrac{6a^2b}{2a}$  h) $\dfrac{6y^4z}{y^3z^2}$

i) $\dfrac{14y^6z^7}{2z^8}$  j) $\dfrac{33x^5yz^9}{11x^9y^4}$  k) $\dfrac{4x^4y^4z^4}{8x^2y^3z^4}$  l) $\dfrac{10a^3b^2c^6}{25a^9b}$

**5** Simplify these.

a) $\dfrac{3x^2}{x^5}$  b) $\dfrac{4x^3y^2}{2xy^4}$  c) $\dfrac{9x^4y^3z^9}{3x^4y^2z^4}$  d) $\dfrac{8a^3b^2c^7}{12abc^5}$

e) $\dfrac{25a^5b^2c}{5ab^2c^3}$  f) $\dfrac{24ab^9c^4}{28a^6b^2c^3}$  g) $\dfrac{14a^2b^2c^2}{21a^3bc^7}$  h) $\dfrac{16a^4b^4c^4}{18a^5bc}$

170

## 15: Fractions in algebra

**Mixed Exercise**

**6** Simplify these.

a) $\dfrac{(x-1)(x+1)}{(x+2)(x-1)}$  b) $\dfrac{(x+4)(x-2)}{(x-2)(x+4)}$  c) $\dfrac{(2x-1)(x-2)}{(x+3)(2x-1)}$

**7** a) Factorise (i) $5x^2 + 2x$  (ii) $6x^3 - 2x^2$

b) Use part a) to simplify (i) $\dfrac{x^3}{5x^2 + 2x}$  (ii) $\dfrac{2x^4}{6x^3 - 2x^2}$

**8** a) Factorise (i) $2x^2 + 2x$  (ii) $3x^2 + 6x$  (iii) $2x^3 + 4x^2$

b) Use part a) to simplify (i) $\dfrac{2x^2 + 2x}{2x}$  (ii) $\dfrac{3x^2 + 6x}{3x}$  (iii) $\dfrac{2x^3 + 4x^2}{2x}$

**9** a) Factorise $x^3 - x^2$

b) Use part a) to simplify $\dfrac{x^3 - x^2}{x - 1}$

**10** a) Factorise (i) $x^2 + x - 2$  (ii) $x^2 + 5x + 6$  (iii) $x^2 + 3x - 4$

b) Use part a) to simplify (i) $\dfrac{x^2 + x - 2}{x + 2}$  (ii) $\dfrac{x^2 + 5x + 6}{x + 2}$  (iii) $\dfrac{x^2 + 3x - 4}{x - 1}$

**11** a) Factorise (i) $x^2 + 5x - 6$  (ii) $x^2 - 3x + 2$

b) Use part a) to simplify $\dfrac{x^2 + 5x - 6}{x^2 - 3x + 2}$

**12** a) Factorise (i) $x^2 + 3x + 2$  (ii) $x^2 + 4x + 3$

b) Use part a) to simplify $\dfrac{x^2 + 3x + 2}{x^2 + 4x + 3}$

**13** a) Factorise (i) $x^2 - 9$  (ii) $x^2 + 5x + 6$

b) Use part a) to simplify $\dfrac{x^2 - 9}{x^2 + 5x + 6}$

# Sixteen

# Enlargement and similarity

**Before you start this chapter you should be able to**

★ recognise and draw an enlargement of a simple shape using a whole number scale factor.

## Reminder

Shape A has been **enlarged** by scale factor 2 to give shape B. All the lines are twice as long but all the angles are the same size.

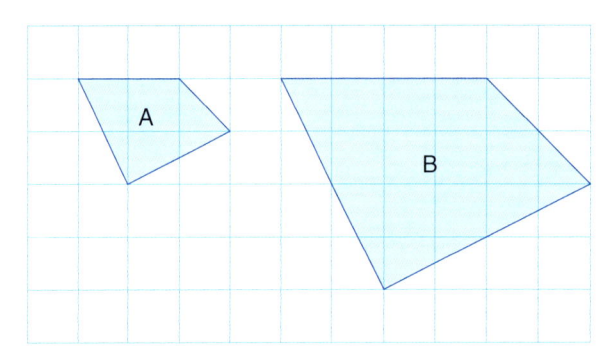

Use the questions in the next exercise to check that you still remember these topics.

### Revision exercise

**1** Which of the following shapes are enlargements of shape A? For those that are, write down the scale factor of the enlargement.

Find the area and perimeter of A, B and C. What do you notice?

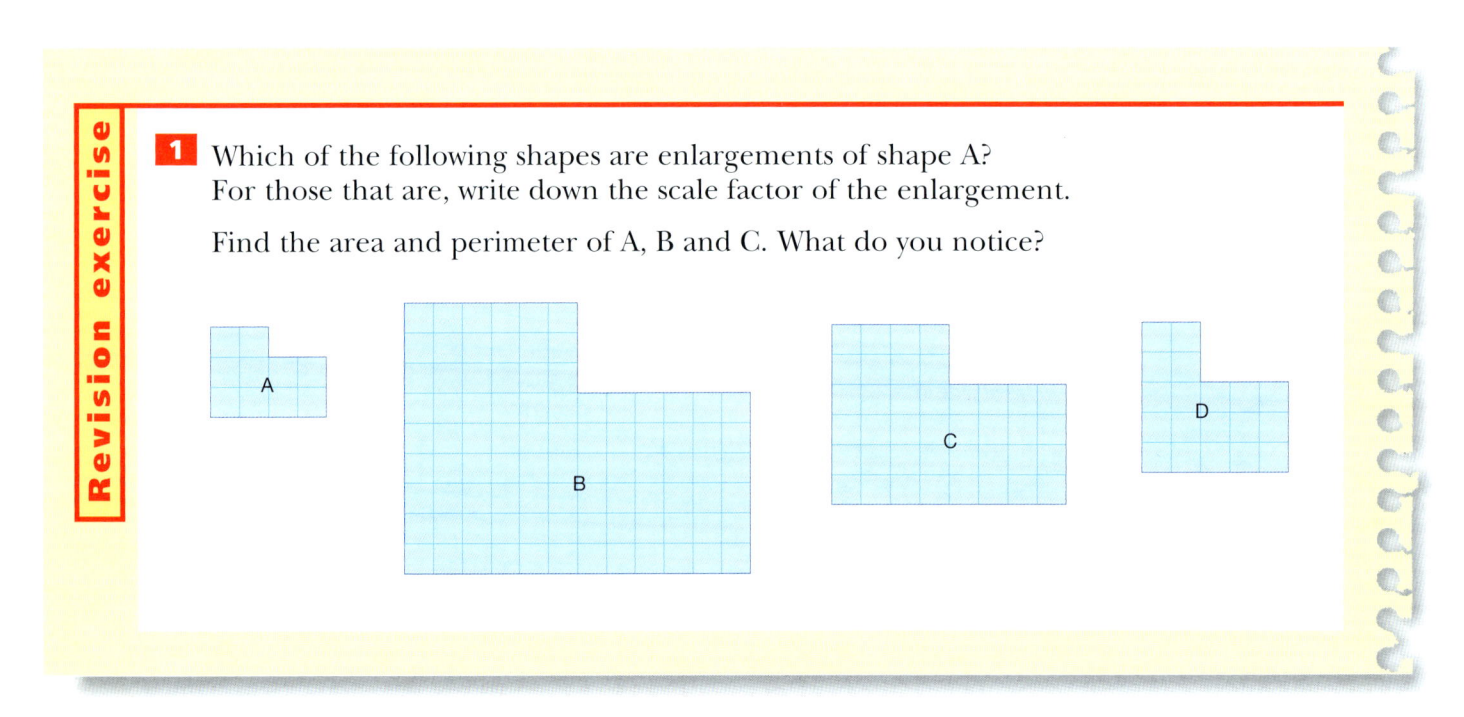

## 16: Enlargement and similarity

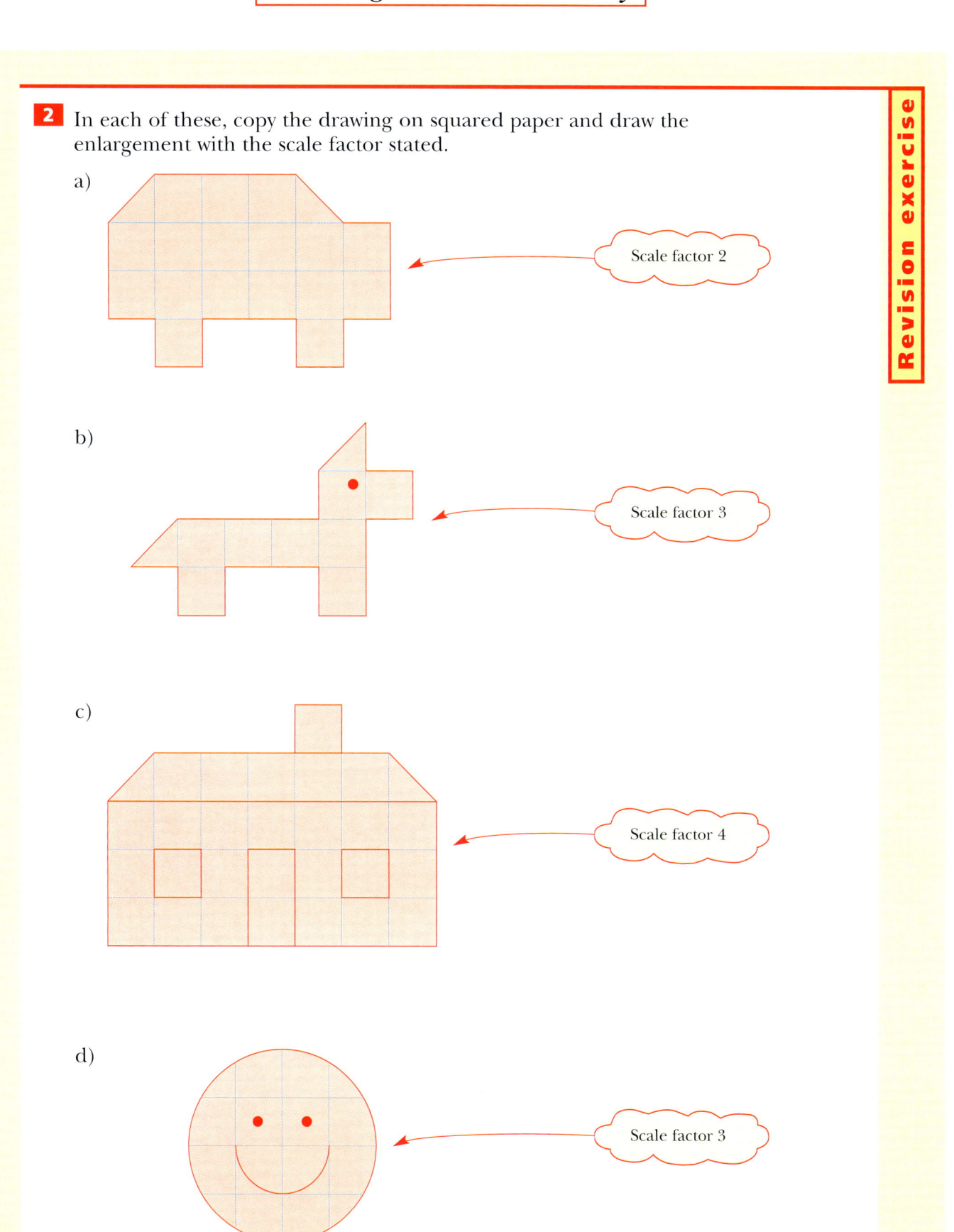

**2** In each of these, copy the drawing on squared paper and draw the enlargement with the scale factor stated.

a) Scale factor 2

b) Scale factor 3

c) Scale factor 4

d) Scale factor 3

*Revision exercise*

173

# 16: Enlargement and similarity

# Centres of enlargement

## Using a centre of enlargement

You can draw enlargements using a centre of enlargement.

This example shows how to draw an enlargement of scale factor 2 using a centre of enlargement (C).

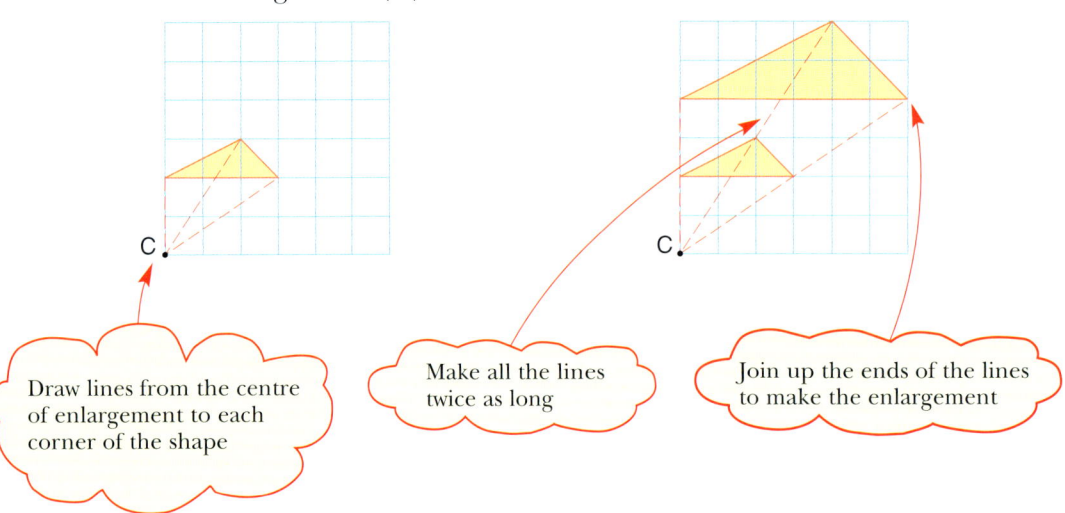

Draw lines from the centre of enlargement to each corner of the shape

Make all the lines twice as long

Join up the ends of the lines to make the enlargement

Check that each side in the enlargement is twice as long as in the original shape.

## Finding a centre of enlargement

You can use the same method in reverse if you are given a shape and its enlargement and you want to find the centre of enlargement and the scale factor.

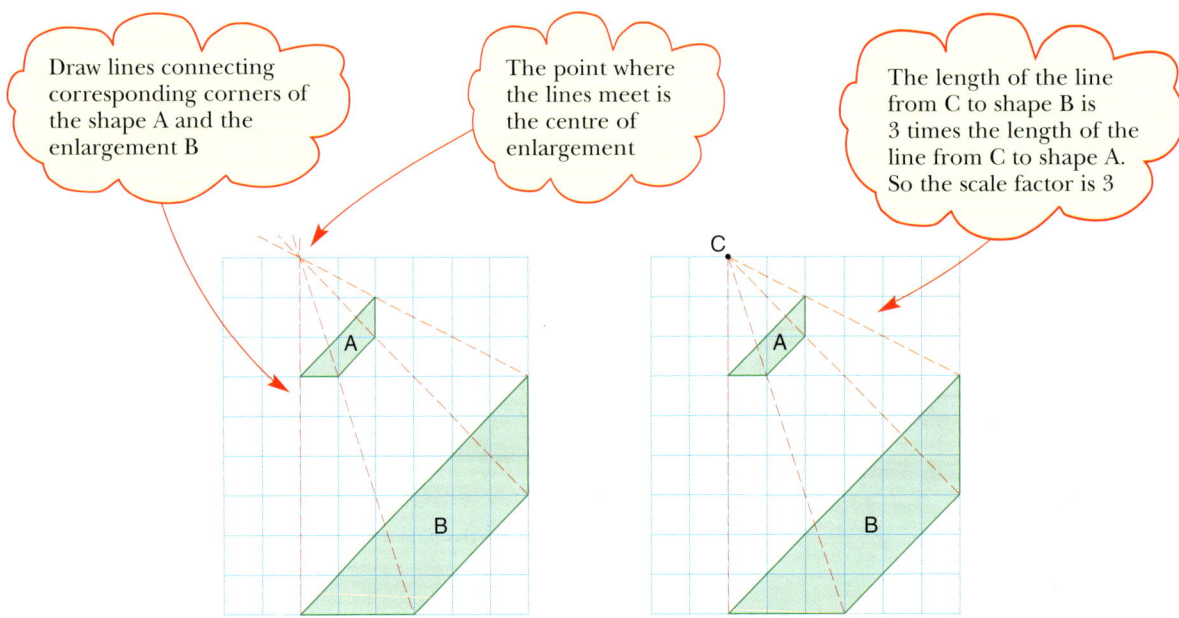

Draw lines connecting corresponding corners of the shape A and the enlargement B

The point where the lines meet is the centre of enlargement

The length of the line from C to shape B is 3 times the length of the line from C to shape A. So the scale factor is 3

# 16: Enlargement and similarity

**1** For each of these, copy and extend the diagram. Draw an enlargement with scale factor 2, using the point C as the centre of enlargement.

a)

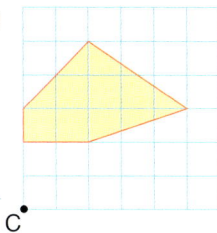

b)

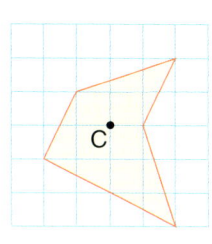

**2** For each of these, copy the diagram and draw an enlargement with scale factor 3, using the point C as the centre of enlargement.

a)

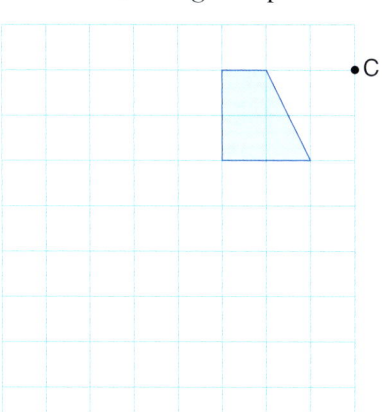

b)

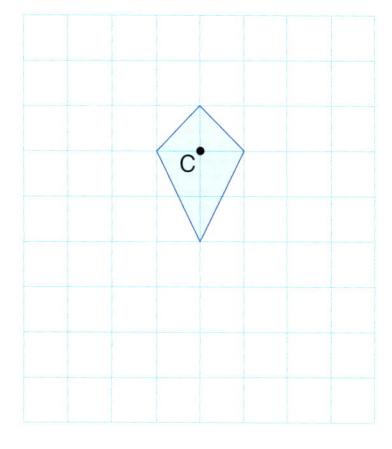

**3** In each of the diagrams below, shape A has been enlarged to make shape B. In each case find the scale factor and the co-ordinates of the centre of enlargement.

a)

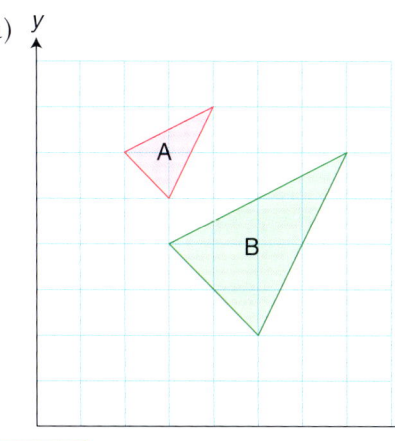

b)

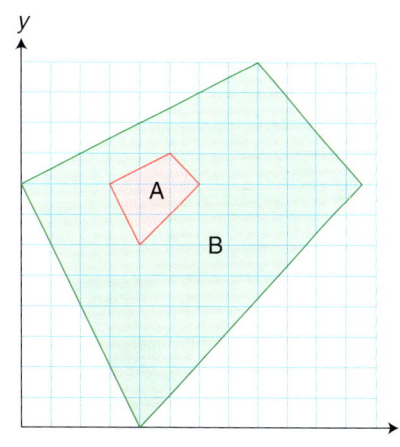

Find out how the zoom facility works on a graphic calculator and write instructions on it for a friend who is younger than you.

## 16: Enlargement and similarity

# Scale factors less than 1

This photograph has been reduced in size to fit into a magazine column. Each side is half as long as it was in the original.

Even though this is a reduction in size, in mathematics it is still called an enlargement. It is an enlargement with scale factor $\frac{1}{2}$.

> **An enlargement with scale factor greater than 1 makes things larger.**
> **An enlargement with positive scale factor less than 1 makes things smaller.**

**?** What do you think a scale factor of $\frac{3}{2}$ means?

You can draw enlargements with fractional scale factors using a centre of enlargement. This example shows how to draw an enlargement with scale factor $\frac{1}{2}$. You can see that the method is the same as before.

*First draw lines from the centre of enlargement, C, to each corner of the shape*

*Then make all the lines half as long*

*Join up the ends of the lines to make the enlargement*

**?** The examples on this page are all two-dimensional.

You often meet enlargements of three-dimensional objects too.

One such enlargement is shown in the photograph. Estimate its scale factor.

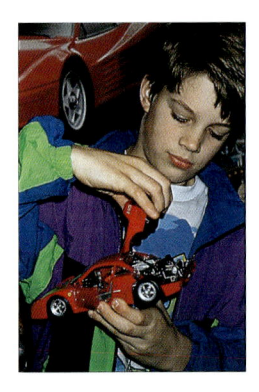

**176**

## 16: Enlargement and similarity

**1** A drawing is 18 cm wide.

Find the width of each of these photocopied enlargements.

a) Scale factor 3
b) Scale factor $\frac{1}{4}$
c) Scale factor $\frac{3}{2}$
d) Scale factor $\frac{2}{3}$

**2** For each of these, copy the diagram and draw an enlargement with the given scale factor and centre of enlargement (C).

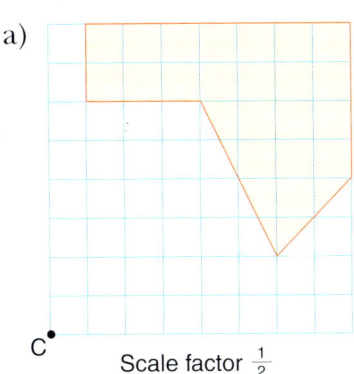

a) Scale factor $\frac{1}{2}$

b) Scale factor $\frac{3}{2}$

c) Scale factor $\frac{3}{4}$

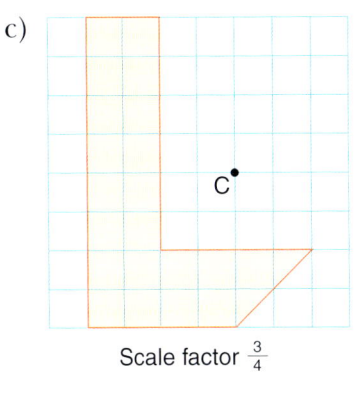

**3** A photograph is 12 cm wide. The widths of some enlargements of the photograph are given below.

Find the scale factor of each enlargement.

a) 24 cm
b) 4 cm
c) 36 cm
d) 6 cm

**4** In each of these, shape A has been enlarged to create shape B. Find the scale factor and the co-ordinates of the centre of enlargement.

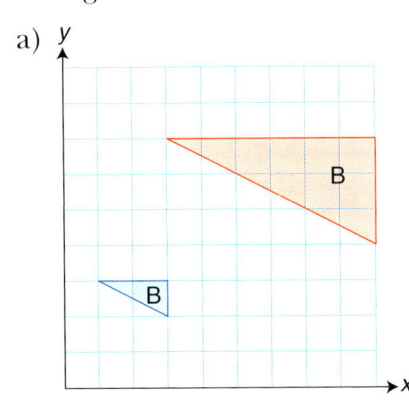

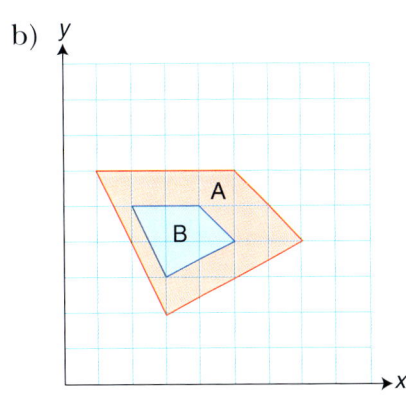

a)

b)

Find 3 examples of enlargements of solid objects.

For each one, work out its scale factor.

177

# 16: Enlargement and similarity

## Similar shapes

In mathematics, the word similar does not just mean that two things are rather alike.

**Similar shapes** are shapes that are enlargements of each other.

Look at this poster that is displayed outside a photographic studio.

There is a mistake in one of the print sizes. They should all be similar shapes.

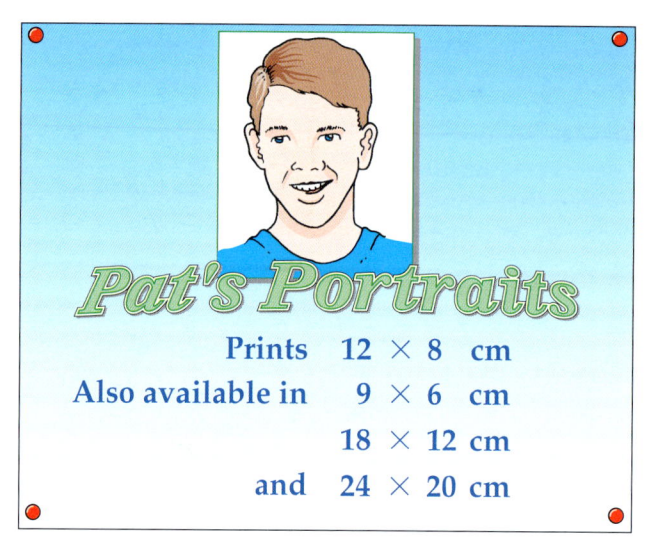

**Pat's Portraits**

Prints 12 × 8 cm
Also available in 9 × 6 cm
18 × 12 cm
and 24 × 20 cm

*Can you see which size is incorrect?*

To check that two shapes are similar you need to make sure that the height and width have been enlarged by the same scale factor.

You find the scale factor for the width by dividing the new width by the old width.

For the 9 × 6 cm print this gives

$$\text{width scale factor} = \frac{6}{8} = 0.75$$

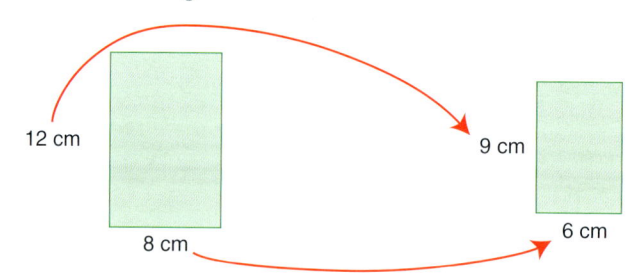

In the same way,

$$\text{height scale factor} = \frac{9}{12} = 0.75$$

The 9 × 6 cm print is a true enlargement because the height and width have been enlarged by the same scale factor.

*Use this method to check which print size is incorrect on the poster.*

## 16: Enlargement and similarity

**Exercise**

**1** Find the scale factor of these enlargements.

a) A photograph 15 cm long is enlarged to 20 cm long
b) A drawing 8 cm wide is enlarged to 18 cm wide
c) A diagram 24 cm long is reduced to 9 cm long.

**2** a) Which of the photographs B – F are similar to photograph A? Explain your answers.

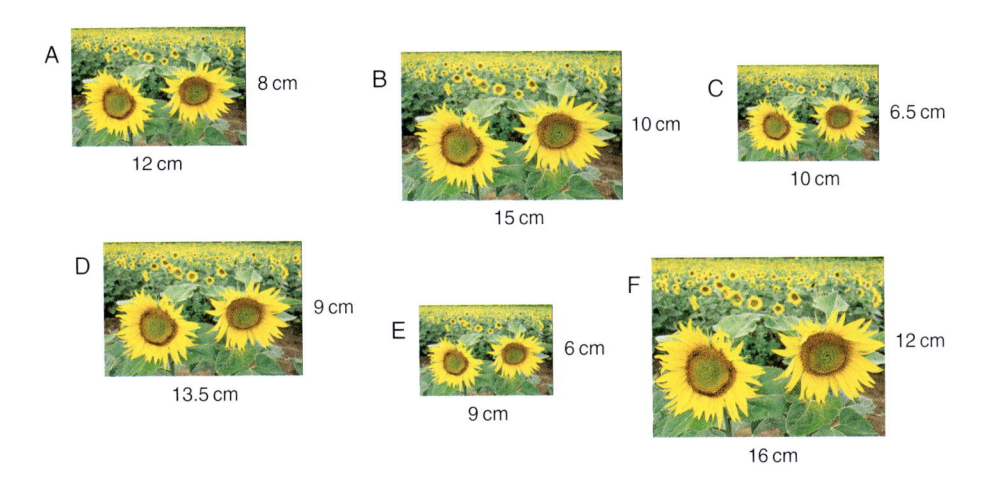

A  8 cm, 12 cm
B  10 cm, 15 cm
C  6.5 cm, 10 cm
D  9 cm, 13.5 cm
E  6 cm, 9 cm
F  12 cm, 16 cm

b) I need an enlargement of photograph A, 20 cm high.

How wide will the enlarged photograph be?

**3** Each of the triangles D, E and F is similar to one of the triangles A, B and C.

Match up the pairs of similar triangles, colouring the equal angles.

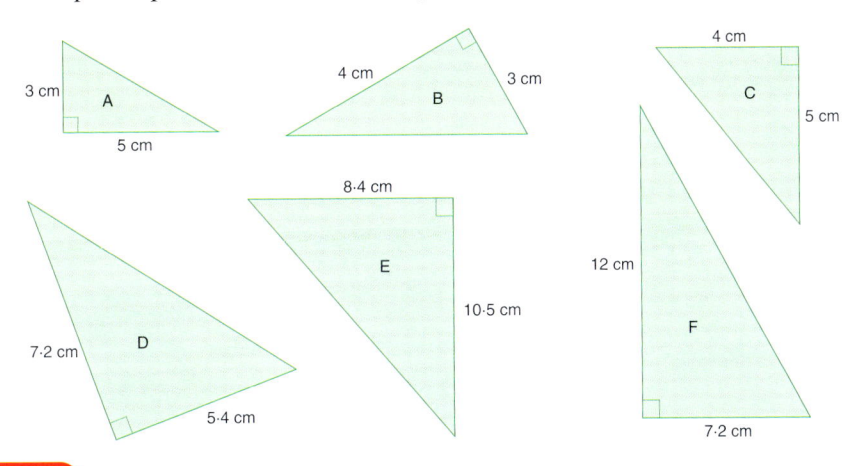

A: 3 cm, 5 cm
B: 4 cm, 3 cm
C: 4 cm, 5 cm
D: 7·2 cm, 5·4 cm
E: 8·4 cm, 10·5 cm
F: 12 cm, 7·2 cm

---

Measure the width and height of at least 3 different television screens. Are they similar to each other?

### 16: Enlargement and similarity

# Using similarity

Many television screens have similar shapes.

**?** *Why do they have similar shapes?*

Justin wants to buy a larger television. His TV has a 34 cm screen and he would like a 51 cm one. Television screen sizes are given by the length of the diagonal, which is usually 34 cm, 48 cm, 51 cm, 59 cm or 66 cm.

First Justin wants to know the width of a 51 cm screen so that he can get an idea of how much space it will take up. He measures his 34 cm screen and finds that it is 29 cm wide.

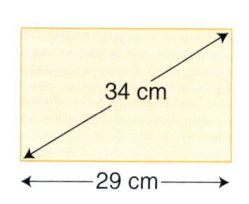

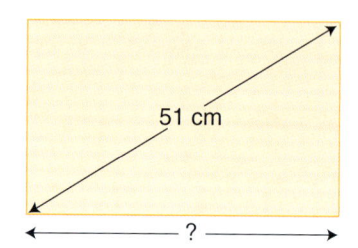

The new screen is an enlargement of Justin's old screen. This means that all the dimensions have been multiplied by the same scale factor.

Justin works out the scale factor like this:

> The scale factor is $\frac{\text{new diag}}{\text{old diag}}$
> $= \frac{51}{34}$
> $= 1.5$

**?** *Check that multiplying 34 by $\frac{51}{34}$ gives 51.*

*What would the scale factor be if Justin wanted a 59 cm screen?*

Justin works out the width of the new screen like this:

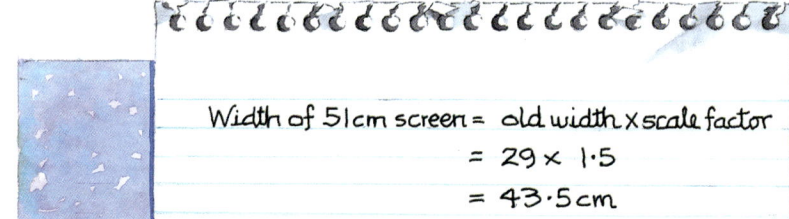

> Width of 51 cm screen = old width × scale factor
> $= 29 \times 1.5$
> $= 43.5$ cm

**The new screen is 43.5 cm wide.**

**?** *The pictures in films made for the cinema are not similar to television pictures. What happens to these films when they are shown on TV?*

# 16: Enlargement and similarity

**Exercise**

**1** In each of these questions there is a pair of similar shapes.

Find the missing lengths in each case. (The angles marked in red and blue are equal.)

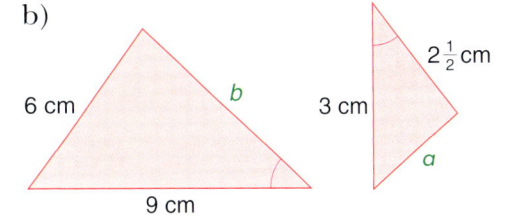

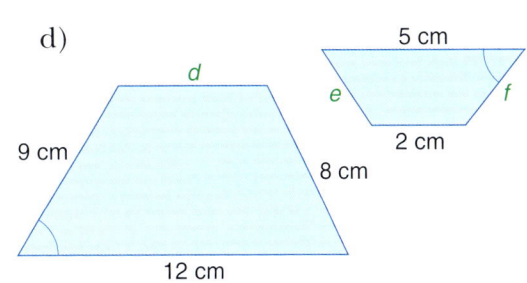

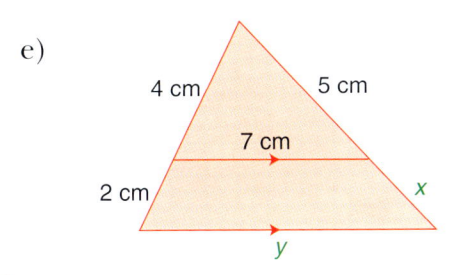

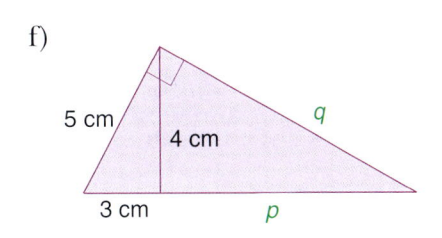

a) [Two similar triangles: larger with sides 3 cm, x, y; smaller with sides 2 cm, 4 cm, 3 cm]

b) [Two similar triangles: larger with sides 6 cm, b, 9 cm; smaller with sides $2\frac{1}{2}$ cm, 3 cm, a]

c) [Two similar quadrilaterals: larger with sides 10 cm, 5 cm, 8 cm; smaller with sides p, q, 4 cm]

d) [Two similar trapezia: larger with sides 9 cm, d, 8 cm, 12 cm; smaller with sides 5 cm, e, 2 cm, f]

e) [Triangle with a line parallel to the base; upper part has sides 4 cm, 5 cm, 7 cm; lower extensions 2 cm, x, y]

f) [Right-angled triangle split by an altitude: sides 5 cm, 4 cm, 3 cm, q, p]

**2** Joanna and her sister Sarah have been on holiday in Canada.

Joanna took a picture of Sarah standing by a Giant Redwood tree (one of the tallest trees in the world).

In the photograph, the tree is 9.9 cm wide and Sarah is 1.2 cm tall. Sarah is 1.1 m tall in reality.

How wide is the tree?

**3** The different 'A' sizes of paper, such as A3, A4 and A5, are all similar shapes. As the numbers go up, the paper sizes get smaller. The height of a particular size of paper is always the same as the width of the previous size. A4 paper is 210 mm wide and 297 mm high.

Work out the dimensions of all paper sizes from A0 (the biggest) to A6. Write your answers in a table like this.

| Paper size | Height (mm) | Width (mm) |
|---|---|---|
| A3 | | |
| A4 | 297 | 210 |
| A5 | 210 | |
| | | |

181

## 16: Enlargement and similarity

# Finishing off

**Now that you have completed this chapter you should**

★ be able to recognise and draw an enlargement of a simple shape using a whole number scale factor and a centre of enlargement

★ be able to recognise and draw an enlargement of a simple shape using a fractional scale factor and a centre of enlargement

★ understand what is meant by mathematical similarity and be able to use it to solve problems.

Use the questions in the next exercise to check that you understand everything.

## Mixed exercise

**1** In each of these, copy the diagram and draw an enlargement with the given scale factor and centre of enlargement (C).

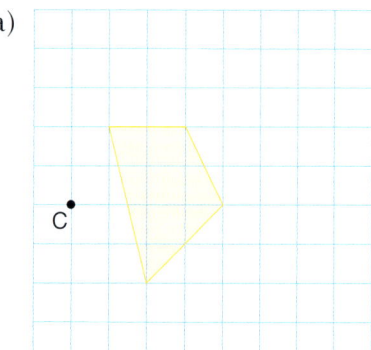

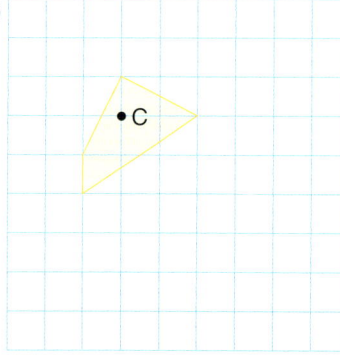

a) Scale factor 2

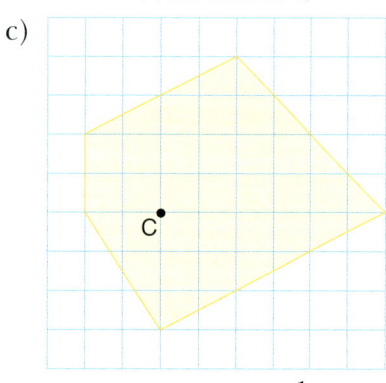

b) Scale factor 3

c) Scale factor $\frac{1}{2}$

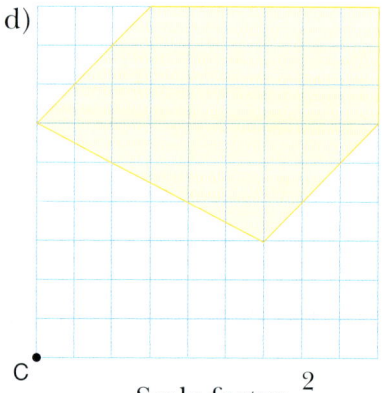

d) Scale factor $\frac{2}{3}$

182

# 16: Enlargement and similarity

**Mixed exercise**

**2** In each of these, shape Q is an enlargement of shape P.

Find the scale factor and the co-ordinates of the centre of enlargement.

a)

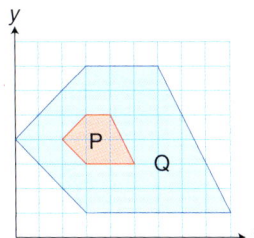

b)

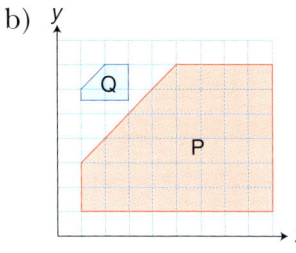

**3** For each of these pairs of similar shapes, find the lengths marked with letters.

a)

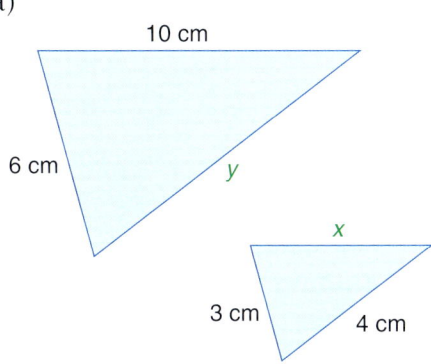

b)

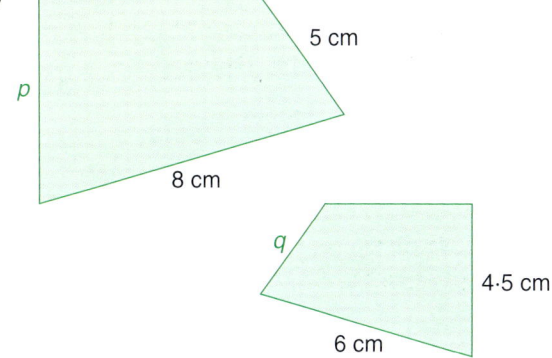

**4** The diagram below shows the cross-section of a roof frame. A horizontal strut AB has been put in for extra strength.

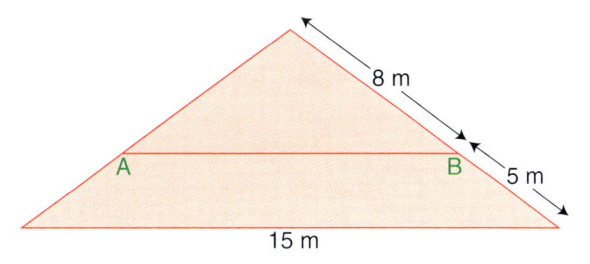

Find the length of the strut AB.

On a sunny day, measure your own height and the length of your shadow.

Now measure the lengths of the shadows of some tall trees or buildings.

Use similarity to work out the heights of the buildings or trees.

183

# Answers

## Chapter 1: Fractions, decimals and percentages

### Pages 4–5: Revision exercise

1. Badminton $\frac{1}{2}$, Football $\frac{1}{4}$, Gym $\frac{1}{8}$, Yoga $\frac{1}{8}$
2. a) $2\frac{3}{4}$  b) $2\frac{1}{6}$  c) $3\frac{5}{8}$  d) $4\frac{2}{3}$  e) $3\frac{1}{7}$  f) $3\frac{3}{5}$
3. a) $\frac{13}{4}$  b) $\frac{31}{8}$  c) $\frac{27}{10}$  d) $\frac{17}{3}$  e) $\frac{37}{5}$  f) $\frac{77}{16}$
4. a) $3\frac{1}{2}$  b) $5\frac{1}{8}$  c) $1\frac{5}{16}$  d) $1\frac{9}{16}$  e) $1\frac{3}{16}$
   f) $6\frac{1}{8}$  g) $2\frac{7}{8}$  h) $4\frac{1}{6}$  i) $2\frac{1}{3}$  j) $7\frac{1}{2}$
   k) $1\frac{1}{2}$  l) $7\frac{17}{20}$
5. a) $\frac{51}{100}, \frac{13}{25}, \frac{27}{50}, \frac{11}{20}$  b) $\sqrt{26}, 5.1, \frac{41}{8}, 5\frac{1}{5}$
6. a) 60  b) 45  c) 40
7. a) 5.12  b) 4.19  c) 1.4  d) 24.48
   e) 1.702  f) 6.5  g) 72.5  h) 30
   i) 45  j) 1.05  k) 24 000  l) 84.1
   m) $0.28\dot{3}$  n) 0.06  o) 0.156  p) 9070
8. a) 31.6  b) 45  c) £490.93
9. a) £30  b) 12%  c) £309.60  d) £332
10. a) Philip $63.\dot{3}$% non-smokers;
    Samit $67.\dot{2}\dot{7}$% non-smokers.

### Page 7: Multiplying fractions

1. a) $\frac{1}{6}$  b) $\frac{3}{16}$  c) $\frac{3}{20}$  d) $\frac{5}{8}$
   e) $\frac{1}{4}$  f) $\frac{3}{5}$  g) $\frac{15}{64}$  h) $7\frac{1}{2}$
2. a) $3\frac{1}{2}$  b) $4\frac{1}{2}$  c) $2\frac{2}{3}$  d) $1\frac{3}{5}$
   e) $12\frac{1}{2}$  f) $9\frac{1}{3}$  g) $3\frac{3}{4}$  h) $7\frac{1}{2}$
3. a) $3\frac{1}{4}$  b) $3\frac{3}{8}$  c) $\frac{7}{8}$  d) $1\frac{3}{4}$
   e) $1\frac{3}{4}$  f) 4  g) $3\frac{3}{4}$  h) $7\frac{7}{8}$
   i) $5\frac{1}{2}$  j) 20  k) $4\frac{13}{16}$  l) 12
4. $27\frac{1}{2}$
5. a) (i) 88 cm  (ii) 176 cm
   b) It has doubled (× 2)
   c) (i) 616 cm$^2$  (ii) 2464 cm$^2$
   d) It has quadrupled (× 4)
6. $36\frac{1}{4}$ lbs

### Page 9: Dividing fractions

1. a) $1\frac{1}{4}$  b) $\frac{1}{10}$  c) $\frac{2}{5}$  d) $\frac{1}{9}$
   e) $\frac{1}{2}$  f) $\frac{3}{4}$  g) $\frac{5}{16}$  h) $\frac{1}{4}$
2. a) 12  b) 6  c) 16  d) 30
   e) $\frac{7}{8}$  f) $1\frac{4}{5}$  g) 10  h) $1\frac{3}{4}$
   i) $\frac{3}{4}$  j) 7  k) $2\frac{1}{2}$  l) $3\frac{3}{5}$
3. a) 10  b) 20
4. 40
5. 28
6. a) 39  b) 26  c) 46

### Page 11: Approximate percentages

1. a) £4000  b) £3822.44  c) 20%  d) 18.8%
2. a) 200  b) Disco  c) 15%  d) $\frac{1}{4}$
3. a) 48 feet by 36 feet  b) 1728 square feet
   c) 44%  d) £38
4. a) £18  b) £17.79  c) £150  d) 10%

### Page 13: Finding the original price

1. £75
2. £25 200
3. a) £852  b) £255.60
4. Lamp £80, sofa £200, Chest of drawers £102
5. a) £408  b) £340
6. £5500

### Page 15: Percentage problems

1. a) £160  b) £127.66.44  c) £59.57
2. a) £12 500  b) £13 520
3. a) West
   b) NW 14182, NE 14947, SW 14554, SE 14785
   c) + 2.3%
4. a) £280 000  b) £189 000  c) £378 000, £408 240

## Chapter 2: Formulae and equations

### Pages 20–21: Revision exercise

1. a) 12  b) 15  c) 18  d) 10  e) 0
   f) 20  g) 0  h) 100  i) 0  j) 28
2. a) −10  b) −2  c) −17  d) −4x  e) m  f) 0
3. a) $3a + 6b$  b) $12 − 6x$  c) $15c$  d) $3n^4$
4. a) −15  b) 4  c) −3  d) −3  e) 4
   f) 13  g) $−6y$  h) $2m$
5. a) $x = 81$  b) $x = 7$  c) $y = 69$  d) $x = 17.4$
   e) $x = 9$  f) $x = −3\frac{1}{2}$

184

# Answers

6. a) $2n - 6$  b) $7a - 7b + 7c$
   c) $20a + 24$  d) $12y - 6z$
7. a) $5n - 1$  b) $5$
8. a) $2x + 17$  b) $17 + y$  c) $1 + 4n$  d) $13 - 3a$
9. a) $y = x^2$ because $y$ increases faster than $x$.
   b) $y = \frac{1}{x}$ because $y$ gets smaller as $x$ gets bigger.
   c) $y = x$ because $x$ and $y$ are always equal.

## Page 23: Making up formulae

1. a) 60, 20, 90, –70
   b) 2
2. a) $m = 12y$
   b) $m = 12y = 120$
   c) $y = \frac{m}{12}$
3. a) $C = 100M$
   b) $M = \frac{C}{100}$
4. a) 10 miles
   b) $m = 10l$
   c) $l = \frac{m}{10}$
5. a) (i) $P = 30A$
      (ii) $\frac{P}{30}$
   b) (i) $C = 2.54I$
      (ii) $I = \frac{C}{2.54}$
6. a) $M = D + 12$
   b) $D = M - 12$

## Page 25: Working with unknowns

1. a) $4n + 8$  b) $5m - 10$  c) $28 + 14x$
   d) $16 - 24y$  e) $11 - 3x$  f) $10n + 10$
   g) $6x$  h) $3x$  i) $25x - 6$
   j) $8 - 2n$  k) $7 - 3a$  l) $6 - 2x$
2. a) $n, n - 1, n + 3, 3$
   b) $n, n + 2, 2n + 4, 2n + 2, n + 1, 1$
   c) $n, 10n, 10n + 2, 30n + 6, 30n, 30$
   d) $n, n - 1, 5n - 5, 5n, 5$
3. a) divide answer by 10
   b) $n, n + 10, 10n + 100, 10n$
4. a) $n, n - 1, 4n - 4, 4n + 4, n + 1$, subtract 1
   b) $n, n + 3, 3n + 9, 3n + 6, n + 2$, subtract 2
   c) $n, n - 1, 10n - 10, 10n - 20, n - 2$, add 2

## Page 27: Using equations to solve problems

1. a) $x = 10$  b) $x = 6$  c) $x = 10$  d) $x = 3$
   e) $x = 3$  f) $x = 4$  g) $x = 3$  h) $x = 0$
   i) $x = 1$
2. a) (i) $12x + 16 = 100$  (ii) $x = 7$
   b) (i) $3w + 6 = 72$  (ii) $w = 22$
   c) (i) $2(3w + w) = 600$  (ii) $w = 75$
   d) (i) $4A + 60 + A = 180$  (ii) $A = 24$
   e) (i) $3y = y + 24$  (ii) $h = 12$

3. a) A-serve: £15
      Beeline: £$(5 + 2x)$
      Comic: £$(9 + 1.5x)$
      When $x = 1$, Beeline is cheapest
      When $x = 10$, A-serve is cheapest
   b) $x = 5$  c) $x = 8$  d) $x = 4$  e) A-serve is cheapest for a total of more than 10 hours otherwise use Beeline.

## Page 29: Using graphs to solve equations

1. a) 7  b) –0.12  c) –15
2. a) [graph]
   b) (i) $x = -2, -0.4, 2.4$  (ii) $x = 2.7$
3. a) [graph]
   b) $\frac{3}{4}$
   c) $x = \frac{3}{4}$
   d) Algebra is better in this case.
4. a) Ask your teacher to check your graph.
   b) 1.62  c) 1.6180

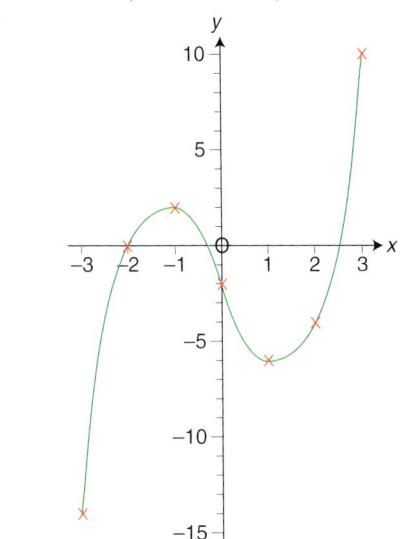

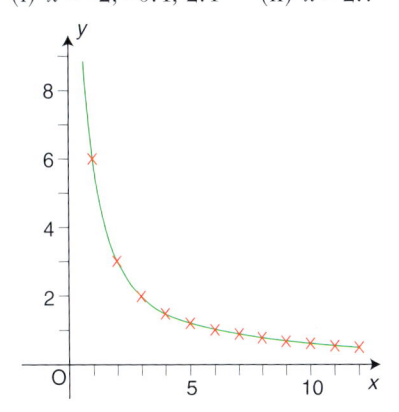

## Page 31: Trial and improvement

1. a) 1 and 2, 1.6  b) 4 and 5, 4.4
   c) 3 and 4, 3.4  d) 9 and 10, 9.0
   e) 4 and 5, 4.8  f) 5 and 6, 5.7

185

# Answers

2  3.141
3  1.325
4  a) $x^2(x+1) = x^3 + x^2$  b) 4 litres = 4000 cm$^3$
   c) $x = 15.5$; 15.5 cm by 15.5 cm by 16.5 cm
5  $(8.87 - 2)$m = 6.87 m

## Page 33: Rearranging a formula

1  a) $x = \dfrac{z-y}{w}$  b) $x = s - t$  c) $x = a(c+b)$

   d) $x = \dfrac{y}{z}$  e) $x = p - qr$  f) $x = s - \dfrac{r}{t}$

   g) $x = \sqrt{cb - a}$  h) $x = \dfrac{e^2 + d}{c}$  i) $x = q(p-r)^2$

   j) $x = h + \sqrt{\dfrac{f}{g}}$  k) $x = \left(\dfrac{w}{z} - y\right)^2$  l) $x = \sqrt{\dfrac{b}{a-c}}$

2  a) $x = \dfrac{d-b}{a-c}$  b) $x = \dfrac{q}{p-q}$  c) $x = \dfrac{y-z}{y+z}$

   d) $x = \dfrac{d}{e-c}$  e) $x = \dfrac{b^2 + ac}{c-b}$  f) $x = \dfrac{p(1+q)}{p-1}$

   g) $x = \dfrac{rst}{t+s}$  h) $x = \dfrac{a^2 + b}{ab}$  i) $x = \dfrac{wz^2 + y}{1 - z^2}$

   j) $x = \sqrt{\dfrac{p}{p^2 - 1}}$

3  $h = \sqrt{\left(\dfrac{A - \pi r^2}{\pi r}\right)^2 - r^2}$

4  b) $r = \sqrt{\dfrac{A}{4 - \pi}}$  c) $r = 5.40$ m

## Chapter 3: Triangles and polygons

### Pages 36–37: Reminder

1  a) $a = 60°$, $b = 70°$, $c = 120°$
   b) $d = 90°$, $e = 100°$,
   c) $f = 110°$
2  a) $a = 82°$
   b) $b = 96°$
   c) $c = 62°$, $d = 30°$
   d) $e = 103°$, $f = 77°$, $g = 103°$
3  a) $a = 111°$, $b = 69°$, $c = 111°$, $d = 111°$, $e = 69°$
   b) $f = 115°$, $g = 65°$, $h = 115°$, $i = 115°$, $j = 65°$,
      $k = 115°$, $l = 65°$, $m = 105°$, $n = 75°$, $o = 105°$,
      $p = 75°$, $q = 105°$, $r = 75°$, $s = 105°$
   c) $t = 50°$, $u = 130°$, $v = 50°$, $w = 130°$, $x = 130°$
4  a) $a = 58°$
   b) $b = 67°$, $c = 113°$, $d = 74°$, $e = 106°$, $f = 74°$
   c) $g = 89°$, $h = 89°$, $i = 47°$
5  a) $a = 40°$
   b) $b = c = d = 25°$
   c) $e = 100°$, $f = g = h = 40°$, $i = 120°$, $j = k = 30°$
   d) $l = 20°$
   e) $m = 85°$, $n = 95°$, $o = 70°$

## Page 39: Angles and triangles

1  a) $a = 70°$
   b) $b = 110°$
   c) $c = 70°$
   d) $d = 80°$, $e = 100°$
   e) $f = 70°$, $g = 60°$
   f) $h = 35°$, $i = 55°$, $j = 15°$
   g) $k = 70°$, $l = 60°$, $m = 50°$
   h) $n = 40°$, $o = 80°$, $p = 100°$, $q = 50°$
2  a) (i) $a = b = 70°$, $x = y = 110°$  (ii) isosceles
   b) (i) $a = b = c = 60°$  (ii) equilateral
   c) (i) $a = 95°$, $b = 70°$, $c = 15°$  (ii) scalene

## Page 41: Drawing triangles

1  6.3 cm, 4.3 cm
   4.5 cm, 7.9 cm
2  a) AC = 5.3 cm, BC = 6.5 cm
   b) PR = 9.2 cm, QR = 6.4 cm
3  a) 94°, 56°, 30°  b) 87°, 30°, 63°
   c) 94°, 36°, 50°  d) 102°, 56°, 22°
5  a) Can  b) Cannot  c) Cannot
   d) Can  e) Cannot

## Page 43: Drawing more triangles

1  a) DF = 6.8 cm
   b) YX = 8.9 cm
2  a) GH = 7.4 cm, GFH = 84°
   b) LM = 3.9 cm, LNH = 43°
4  b) $p = 85°$  $q = 128°$  $r = 23°$  $s = 70°$
   c) $a = 3.8$ m  $b = 6.0$ m
      $c = 6.4$ m  $d = 10.9$ m

## Page 45: Polygons

1  A — regular pentagon
   B — irregular octagon
   C — irregular hexagon
   D — irregular quadrilateral
   E — irregular pentagon
   F — regular octagon
2  a) 120°  b) 90°  c) 72°
   d) 60°  e) 45°  f) 36°
3  144°

## Page 47: Angle sum of a polygon

1  a) 720°  b) 1080°  c) 1260°  d) 1440°
2  a) 120°  b) 135°  c) 140°  d) 144°
3  a) 60°  b) 45°  c) 40°  d) 36°
4  a) Hexagon (120°)  b) Octagon (135°)
   c) Nonagon (140°)  d) Decagon (144°)
5  a) 12  b) 9  c) 6  d) 4
   e) 10  f) 5

# Answers

## Chapter 4: Graphs and co-ordinates

### Page 51: Working with co-ordinates

1. a) (2, 6)  b) (7, 5)  c) (7, 2)  d) (6, 4)
   e) (7, 6)  f) (2, 8)
2. (5, 2)
3. B (2, 4), D (5, 8)
4. B (5, 11), D (9, 17)
5. (10, 11)
6. a) (0, 0, 0)  b) (0, 0, 2)  c) (1, 0, 0)  d) (3, 0, 0)
   e) (2, 1, 0)  f) (2, 1, 3)  g) (4, 0, 2)  h) (3, 1, 3)
   i) (0, 1, 2)
7. C (6, 0, 0), D (9, 3, 12), E (2, 2, 9), F (9, 9, 12)

### Page 53: Using co-ordinates

1. (3, 0), (−1, 2) or (−5, −4)
2. (0, 3), (8, −1) or (−12, −5)
3. a) (4, 2)  b) 5
4. a) (5, 9)  b) 10, 10  c) rectangle
5. a) (−5, −6)  b) $\sqrt{80}$, $\sqrt{80}$  c) rhombus
6. a) (3, 4)  b) $\sqrt{40}$, $\sqrt{40}$  c) $\sqrt{80}$, $\sqrt{80}$  d) square
7. 13
8. a) AB = BC = $\sqrt{20}$  b) AB = AC = 25

### Page 55: Graphs for real life

1. b) (i) Lights Galore  (ii) Shadows
2. b) 1000 miles  c) Worry-Free, £24
3. b) (i) Cheap-Chatter  (ii) Teletalk
   (iii) Fone-a-Friend
4. b) (i) £200  (ii) £380  (iii) £540

### Page 57: Number patterns and sequences

1. a) 1 2 3 4 5 6 7 8 | $n$
      2 4 6 8 10 12 14 16 | $2n$
   b) 1 2 3 4 5 6 7 8 | $n$
      10 12 14 16 18 20 22 24 | $2n + 8$
   c) 1 2 3 4 5 6 7 8 | $n$
      1 3 5 7 9 11 13 15 | $2n - 1$
   d) 1 2 3 4 5 6 7 8 | $n$
      0 3 6 9 12 15 18 21 | $3n - 3$
   e) 1 2 3 4 5 6 7 8 | $n$
      1 4 9 16 25 36 49 64 | $n^2$
   f) 1    2    3    4    5    6    7
      1×2, 2×3, 3×4, 4×5, 5×6, 6×7, 7×8,
      8 | $n$
      8×9 | $n(n+1)$

2. a) $1 + 3 + 5 + 7 + 9 = 25$
      $1 + 3 + 5 + 7 + 9 + 11 = 36$
   b) 100, $n^2$
3. a) (i) £12m  (ii) £6m  (iii) £4m
      (iv) £3m  (v) £2.4m  (vi) £$\frac{12}{n}$m
   b) (i) £12m  (ii) £8m  (iii) £6m
      (iv) £4.8m  (v) £4m  (vi) $\frac{£24}{(m+1)}$
   c) Different
4. a) $5 \times 4 + 5, 6 \times 5 + 6$
   b) $8 \times 7 + 8$
      $n(n-1)+n$
   c) 1, 4, 9, 16, 25, 36, 64, $n^2$
   d) $n(n-1) + n$
      $= n^2 - n + n$
      $= n^2$

## Chapter 5: Simultaneous equations

### Page 61: Using simultaneous equations

1. a) $x = 7$  b) $14 + y = 15$ so $y = 1$  c) $7 + 1 = 8$
2. a) 4, 9  b) 10, 1  c) 6, −1  d) 2, 2
   e) 5, 3  f) 6, 1  g) 1, 5  h) 100, 200
   i) −2, 3
3. a) 7, 3  b) 3, $\frac{1}{3}$  c) 9, 11  d) 1, 0
   e) 2, −1  f) 5, $\frac{1}{2}$
4. White costs £2, blue £2.50
5. Oranges cost 40p, apples cost 30p

### Page 63: More simultaneous equations

1. a) 10, 2  b) 18, 3  c) 8, 5
   d) 9, 5  e) 7, −1  f) $11\frac{1}{2}$, $1\frac{1}{2}$
2. a) $20x - 4y = 208$  b) $23x = 230$  c) 10, −2
3. a) 20, 11  b) $\frac{1}{2}$, 1  c) 10, 5  d) 5, 2  e) 5, 1
   f) $1\frac{1}{2}$, 4  g) 8, $3\frac{1}{2}$  h) 10, 2  i) $\frac{1}{2}$, $\frac{1}{4}$
4. Jam doughnuts cost 18p, ring doughnuts 12p.
   She could buy 11 (with 2p change).
5. Carl is 34, his father 66 years old.

### Page 65: Multiplying both equations

1. a) $6x + 10y = 580$  b) $25x - 10y = 350$
   c) $31x = 930$  d) 30, 40
2. a) 9, 2  b) 8, 0  c) 10, 2  d) 5, −1  e) 0, 5
   f) $\frac{1}{2}$, 1  g) 9, 7  h) $\frac{1}{2}$, $-\frac{1}{3}$
3. 5 trays of barfi, 3 of halwa
4. Basic rate is £3.50, rate after 11 p.m. is £6 per hour.
5. Carnations cost 50p, roses 90p.

# Answers

## Page 67: Other methods of solution

1. a) 4, 6   b) 6, 10   c) 4, 3   d) 5, 1
   e) 6, –5   f) 1, 2
2. a) $x + 2y = 12$   b) $x + y = 8$   c) 4, 4
3. 5 in a packet, 20 in a box
4. a) 35, 71   b) 2, 5   c) $2\frac{1}{2}, \frac{1}{2}$   d) –1, 5
   e) 1, –1   f) 6, 1

# Chapter 6: Trigonometry

## Page 71: Introduction to trigonometry

1. 

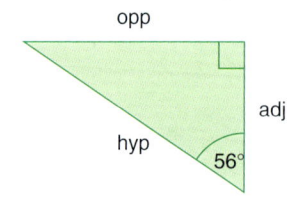

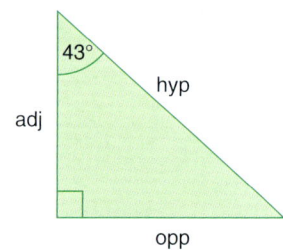

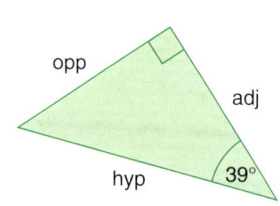

Your answers to questions 2 and 3 may be slightly different (up to about 0.05 either way) from those given. This is because it is not possible to measure accurately.

2. b) 0.84   c) 0.64   d) 0.77
3. b) 1.33   b) 0.80   d) 0.60

## Page 73: Using tangent (tan)

1. a) 0.466   b) 3.271   c) 1   d) 0.754
   e) 1.235   f) 7.115
2. a) 3.63 cm   b) 4.43 cm   c) 5.60 cm   d) 3.78 cm
   e) 9.20 cm   f) 2.98 cm
3. 33.7 m

## Page 75: Finding the adjacent side using tan

1. $a$ = 3.90 cm   $b$ = 9.40 cm   $c$ = 5.78 cm
2. $d$ = 60.0°   $e$ = 66.9°   $f$ = 47.3°
3. $p$ = 4.40 cm   $q$ = 46.3°   $r$ = 4.93 cm
   $s$ = 28.7°   $t$ = 3.56 cm   $u$ = 3.08 cm

## Page 77: Using sine (sin)

1. $a$ = 4.46 cm   $b$ = 8.99 cm   $c$ = 6.19 cm   $d$ = 8.04 cm
   $e$ = 5.00 cm   $f$ = 7.12 cm
2. $p$ = 50.2°   $q$ = 37.3°   $r$ = 37.6°
3. a) 7.73 m   b) 48.6°

## Page 79: Using cosine (cos)

1. $a$ = 4.82 cm   $b$ = 8.40 cm   $c$ = 6.87 cm   $d$ = 5.35 cm
   $e$ = 8.65 cm   $f$ = 6.48 cm
2. $x$ = 57.6°   $y$ = 33.3°   $z$ = 59.0°
3. 2.008 km

## Page 81: Using sin, cos and tan

1. $a$ = 6.86 cm   $b$ = 4.17 cm   $c$ = 9.85 cm   $d$ = 5.70 cm
   $e$ = 8.49 cm   $f$ = 3.96 cm   $g$ = 3.87 cm   $h$ = 8.48 cm
   $i$ = 9.12 cm
2. $p$ = 53.6°   $q$ = 37.3°   $r$ = 39.1°   $s$ = 51.5°
   $t$ = 48.7°   $u$ = 60.3°

## Page 83: Using trigonometry

1. a) 127 km south, 272 km west
   b) 409 km north, 368 km west
2. 1067 m
3. a) 932 m   b) 12.5°
4. a) 2.12 m   b) 1.5 m
5. 6.9 m
6. a) 84.5 km south, 181.3 km east
   b) 96.4 km south, 114.9 km west
   c) 180.9 km south, 66.4 km east
   d) 340°

# Chapter 7: Inequalities

## Page 87: Using inequalities

1. a) $x > 3$   b) $x \geq 0$   c) $x < 5$   d) $x \leq -2$
2. a) $x$ is less than 8   b) $p$ is greater than 100
   c) $q$ is greater than or equal to 100
   d) $y$ is less than 17
   e) $x$ is less than or equal to 20
   f) $b$ is greater than or equal to –3
3. a) $2 < 9$   b) $13 > 3$   c) $-3 > -13$   d) $13 > -3$
   e) $3.8 > 3.3$   f) $0.5 < 0.625$
4. a) $s \leq 30$   b) $f \geq 79$   c) $g \geq 20$   d) $p \leq 5$
   e) $c < 250$   f) $n \geq 50$   g) $w \leq 2.5$   h) $d \geq 500$
5. a) $9 > 2$   b) $3 < 13$   c) $13 > -3$   d) $-3 > -13$
   e) $3.3 < 3.8$   f) $0.625 > 0.5$

## Page 89: Number lines

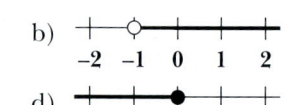

1. a), b), c), 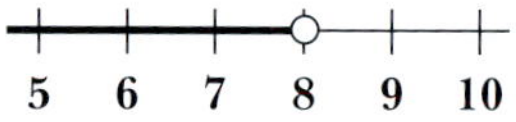 d)

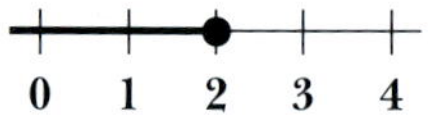

188

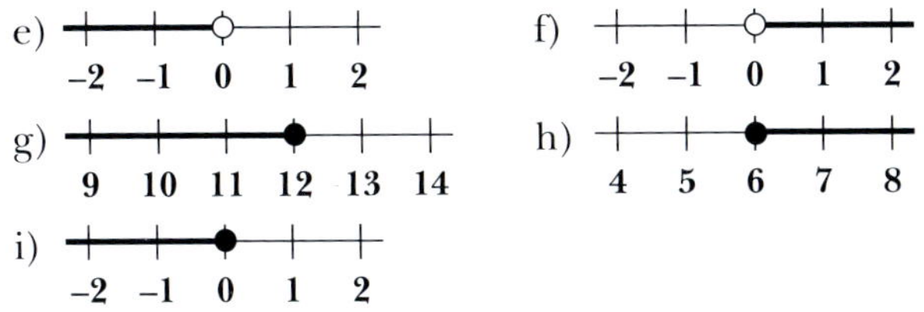

# Answers

e) number line: −2 −1 0 1 2 (open circle at 0, shaded left)
f) number line: −2 −1 0 1 2 (open circle at 0, shaded right)
g) number line: 9 10 11 12 13 14 (shaded 11 to 12)
h) number line: 4 5 6 7 8 (closed circle at 6)
i) number line: −2 −1 0 1 2 (shaded up to closed circle at 0)

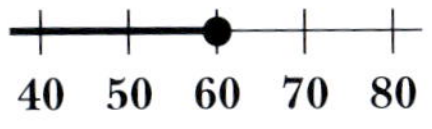

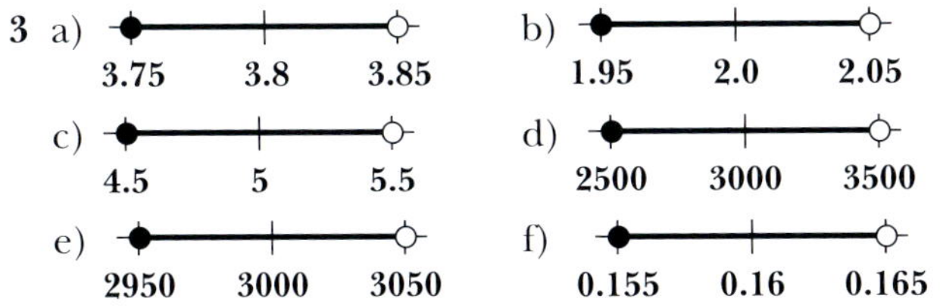

2 a) $w \leq 60$  b) number line: 40 50 60 70 80 (shaded to closed 70)

3 a) 3.75 3.8 3.85 (closed to open)
  b) 1.95 2.0 2.05 (closed to closed)
  c) 4.5 5 5.5 (closed to closed)
  d) 2500 3000 3500 (closed to closed)
  e) 2950 3000 3050 (closed to open)
  f) 0.155 0.16 0.165 (closed to closed)

4 a) 2 3 4 5 6 7 8 (closed 3 to closed 7)
  b) −3 −2 −1 0 1 2 3 4 5 6 (closed −2 to closed 5)
  c) −5 −4 −3 −2 −1 0 1 (closed −4 to open −1)
  d) 1.65 1.7 1.75 (closed to open)
  e) −2 −1 0 1 (open −1 to closed 0)
  f) −7 −6 −5 −4 −3 −2 −1 (open −6 to closed −3)
  g) −1 0 1 2 3 4 (closed 2 to 4)
  h) 0 1 2 3 4 5 6 7 (closed 2 to closed 5)

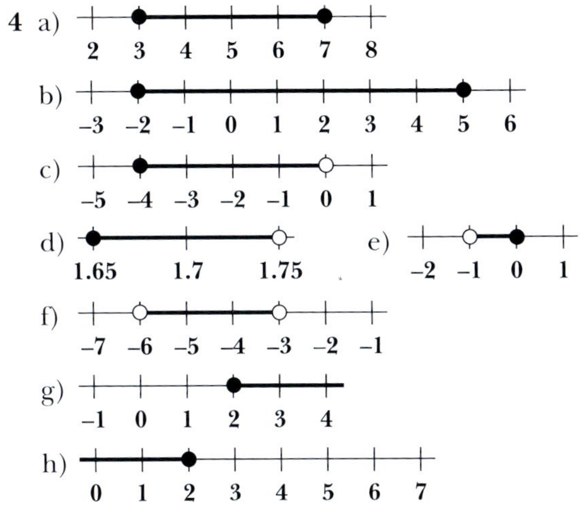

5 $x < 10$, 1, 2, 3, 4, 5, 6, 7, 8, 9
  (number line: asterisks at 1–9)

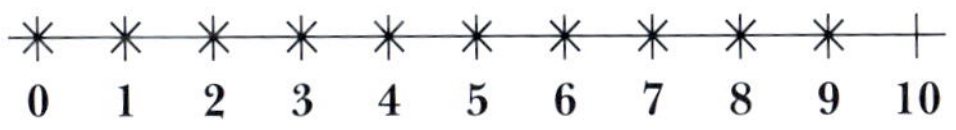

6 a)

| $x$ | −5 | −4 | −3 | −2 | −1 | 0 | 1 | 2 | 3 | 4 | 5 |
|---|---|---|---|---|---|---|---|---|---|---|---|
| $x^2$ | 25 | 16 | 9 | 4 | 1 | 0 | 1 | 4 | 9 | 16 | 25 |

  b) (i) −5 to 4 (closed)
     (ii) −4 to 3 (closed)
     (iii) −10 −5 0 5 10 (open to open)
     (iv) −2 −1 0 1 2 (open to open)
     (v) −4 −3 −2 −1 0 1 2 3 4 (closed to open)

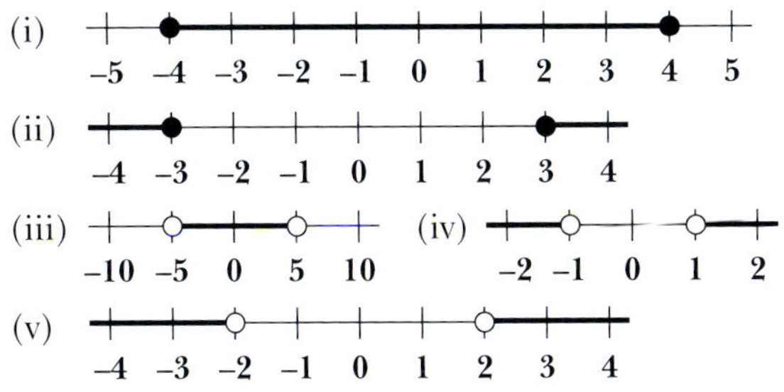

## Page 91: Solving inequalities

1 a) $x < 6$  b) $x \leq 5$  c) $x \geq 8$  d) $x \geq 4$  e) $x < 3$
  f) $x > 6$  g) $x \leq 3.6$  h) $x \leq 8$  i) $x > 2$  j) $x \leq 2$
  k) $x \geq 3$  l) $x \leq -3$

2 a) 2, 3, 5, 7, 11, 13, 17, 19
  b) 25, 36   c) 4, 6   d) 6, 9

3 a) $5x - 200$   b) $5x - 200 > x$; $x > 50$

4 $w + 52 > 14w$, $w < 4$

5 a) $x \geq 11$  b) $x \geq 2$  c) $x > 7$  d) $x < 6$
  e) $1 \leq x \leq 8$  f) $10 \geq x > 3$  g) $6\frac{1}{2} < x < 9\frac{1}{2}$
  h) $2 \leq x \leq 20$  i) $2 < x \leq 11$  f) $x > 11$
  k) $x \geq 16$  l) $x > 1$

## Page 93: Inequalities and graphs

1 a) $-1 \leq x \leq 2, 0 \leq y \leq 2$  b) $x > 2, y > 1$
  c) $0 \leq x \leq 3, 0 \leq y \leq 1$  d) $3 < x < 10, 5 \leq y \leq 10$
  e) $x \geq 4, y \geq 3$  f) $1 \leq x \leq 4, y > -1$

2 Ask your teacher to check your graphs.

3 a) $20 \leq t \leq 120$, $t$ is time in minutes
     $5 \leq d \leq 100$
  b) graph of distance (miles) vs time (minutes), shaded region 20 ≤ t ≤ 120, 5 ≤ d ≤ 100

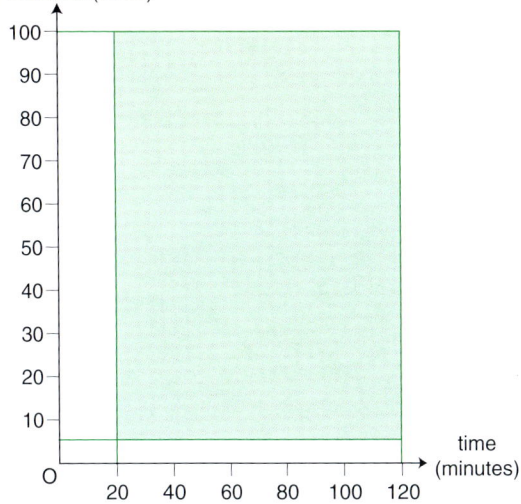

## Page 95: Regions bounded by sloping lines

1 a), b) + d) graph showing points A, B, C, D, E, F with shaded triangular region

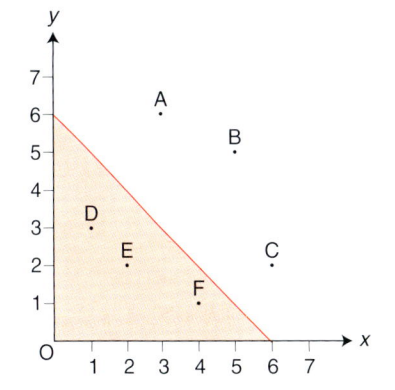

  c)  point   $x + y$
      A       9
      B       10
      C       8
      D       4
      E       4
      F       5

189

# Answers

**2** a), b) + d)

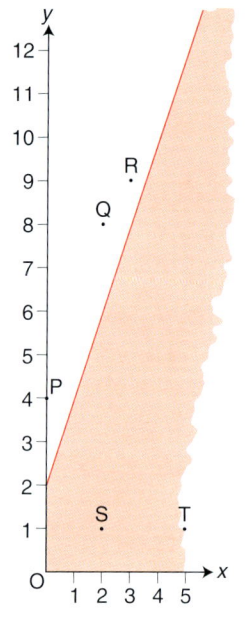

c)
| point | $2x + 2$ |
|---|---|
| P | 2 |
| Q | 6 |
| R | 8 |
| S | 6 |
| T | 12 |

**3** a), b) + c)

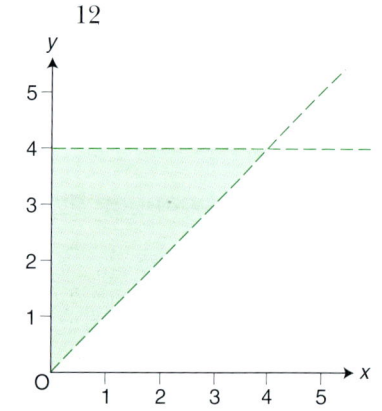

d) Points on the lines are not included.

**4** a) $2x + 2y$
   b) The perimeter can't be longer than the 500 cm strip.
   c) $x > 80$
   d) Ask your teacher to check your graph.

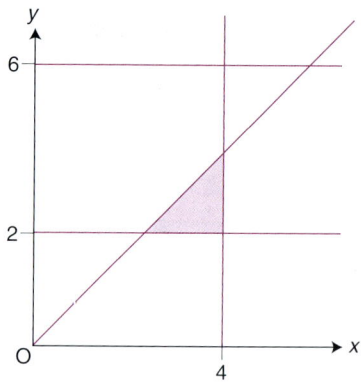

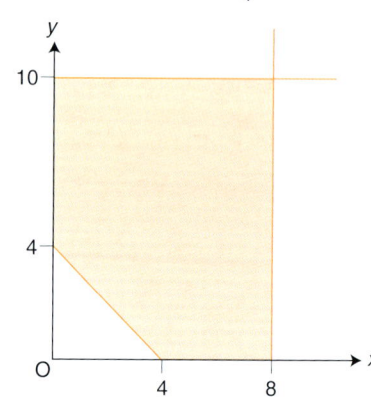

**5** a)

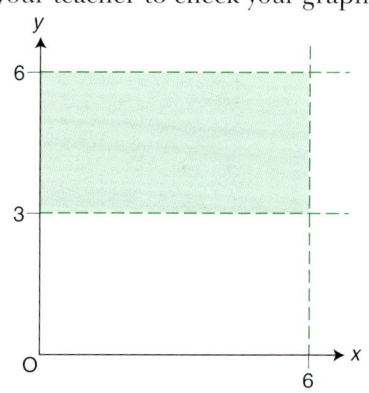

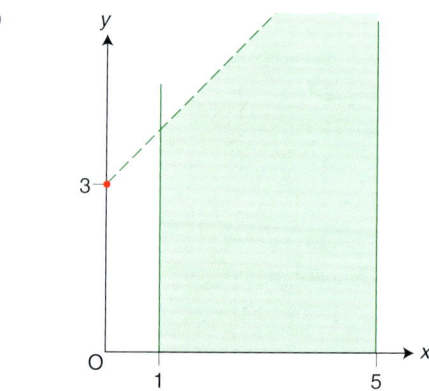

## Page 97: Solution sets

**1** a), c) + d) Ask your teacher to check your graph.
   b) At $(1, 2)$ $x + 3y = 7$
      $(2, 4)$ $x + 3y = 12$

**2** a) $2w + 4s \geq 10$ or $w + 2s \geq 5$
   b) $w \leq 5$
   c) $s \leq 2$
   d) Ask your teacher to check your graph.

**3** a) $x + y \leq 21$   b) $4x + 7y \geq 56$
      Ask your teacher to check your graph.
   c), d) Ask your teacher to check your graph.
   e) $(10\frac{1}{2}, 10\frac{1}{2})$   f) £114

# Answers

## Chapter 8: Indices and standard form

### Pages 100–101: Revision exercise

1. a) 8   b) 36   c) 81   d) 64   e) 243
   f) 256   g) 121   h) 3.375
2. a) 81   b) 9   c) 125   d) 5   e) 8000
   f) 7   g) 10   h) 160 000
3. a) 64   b) 512
4. a) 10   b) 14 by 14   c) 4
5. a) 520   b) 8.3   c) 0.69   d) 2300   e) 0.47
   f) 64   g) 7   h) 5.28   i) 9345   j) 8
   k) 0.912   l) 573 000   m) 0.075   n) 393.71
   o) 140.65   p) 10 560
6. a) 280 000 000 000   b) 57 200 000 000 000

### Page 103: Rules of indices

1. a) $\frac{1}{16}$   b) $\frac{1}{1000}$   c) $\frac{1}{25}$   d) $\frac{1}{8}$   e) $\frac{1}{27}$   f) $\frac{1}{36}$
   g) $\frac{1}{100}$   h) $\frac{1}{81}$   i) $\frac{1}{4}$   j) $\frac{1}{16}$   k) $\frac{1}{216}$
   l) $\frac{1}{10\,000}$
2. a) 49   b) $\frac{1}{9}$   c) $\frac{1}{10}$   d) 64   e) $\frac{1}{81}$   f) 1
   g) 32   h) $\frac{1}{125}$   i) 1 000 000   j) 9   k) $\frac{1}{5}$
   l) 1   m) 216   n) $\frac{1}{64}$   o) 4   p) 625
3. a) $5^6$   b) $2^{11}$   c) $6^4$   d) $10^3$   e) $2^7$   f) $10^6$
   g) $3^8$   h) $4^2$   i) $10^4$   j) $5^{-2}$   k) $3^3$   l) 2
   m) $4^2$   n) $3^{-2}$   o) $2^0$   p) $10^3$

### Page 105: Calculations using standard form

1. a) $8.4 \times 10^4$   b) $6.8 \times 10^6$   c) $1.5 \times 10^{12}$
   d) $1.6 \times 10^7$   e) $1.5 \times 10^9$   f) $7.02 \times 10^{-2}$
   g) $1.44 \times 10^{-17}$   h) $1.6 \times 10^{15}$
2. Mercury $7.86 \times 10^{22}$ kg
   Venus $4.88 \times 10^{24}$ kg
   Mars $2.05 \times 10^{23}$ kg
   Jupiter $2.04 \times 10^{27}$ kg
3. $1.8 \times 10^{-16}$ J
4. a) Europe   b) 9 million   c) 30 million km$^2$
5. a) $2.06 \times 10^{19}$ m$^3$   b) 3592 kg/m$^3$

## Chapter 9: Circles and tangents

### Page 109: Shapes in a circle

1. a) (i) $x = 110°$, $y = 35°$, $z = 55°$ (ii) ACB = 90°
   c) The angle ACB is always 90°.
2. $a = 36°$, $b = 45°$

3. a) $a = 90°$   b) $b = 24°$   c) $c = 102°$
   d) $d = 141°$   e) $e = 68°$   f) $f = 98°$, $g = 72°$
   g) $h = 125°$   h) $i = 62°$
   i) $j = 86°$, $k = 86°$, $l = 94°$   j) $m = 88°$

### Page 111: Angles in a circle

1. a) OA = OC   b) 30°   c) 120°   d) 60°
   e) 80°   f) (i) 140° (ii) 70°
   g) $\angle ACB = \frac{1}{2} \angle AOB$
2. a) $a = 72°$   b) $b = 42°$   c) $c = 64°$   d) $d = 28°$
   e) $e = 35°$   f) $f = 88°$   g) $g = 110°$   h) $h = 70°$
   i) $i = j = k = 25°$   j) $l = 51°$   k) $m = 40°$   l) $n = 25°$
   m) $o = 28°$, $p = 124°$, $q = 28°$   n) $r = 48°$

### Page 113: Tangents

1. a) $a = 36°$   b) $b = 67°$   c) $c = 70°$, $d = 40°$
   d) $e = 64°$   e) $f = 58°$   f) $g = 70°$
2. a) $a = 51°$, $b = 25.5°$   b) $c = 48°$, $d = 42°$, $e = 96°$
   c) $f = 12°$
3. 15 cm
4. 12 cm
5. 8 cm

## Chapter 10: Manipulating expressions

### Page 117: Like terms

1. a) Yes   b) Yes   c) No   d) No   e) No
   f) Yes
2. a) $2x^2 + 3x^2 = 5x^2$, $4x + 7x = 11x$
   b) $y^2 + 2y^2 - y^2 = 2y^2$, $3y + y = 4y$
   c) $3u^2 + 6u^2 = 9u^2$, $4u - 2u + u - 2u = u$
   d) $3p + 6p + 6p = 15p$, $-3p^2 + 5p^2 = 2p^2$, $3 + 2 - 6 = -1$
3. a) $8x + 3y$   b) $10a + 10b$   c) $3p + 4q$
   d) $10s + 9t$   e) $5l + 13m$   f) $5h$
   g) $4x + 3$   h) $2x - 2y + 3$   i) $6a - 6c$
   j) $8p + 11$
4. a) $x^2 + 2x + 1$   b) $x^2 - 2x - 3$   c) $y^2 - 3y + 2$
   d) $2y^2 - 2y + 7$   e) $8a^2 - a$   f) $7 - 5x + x^2$
   g) $-2d$   h) $10x^2 - x + 2$   i) $12x - 5x^2$
   j) $t^2 + 9$
5. a) $3m + 2m^2$   b) $3k + 2k^2$   c) $20x + 2x^2$
   d) $4y$   e) $x^2 - 2x - 8$   f) $a^2 + 4a + 4$
   g) $h^2 - 4h + 5$   h) $h^2 - 9h + 18$   i) $x^2 - 2x^3$

### Page 119: Factorising

1. a) 4   b) 3   c) $a$   d) 5, $b$, 5$b$
2. a) $12 = 3 \times 4$   $3c = 3 \times c$
   b) $16 = 8 \times 2$   $8y = 8 \times y$
   c) $n = n \times 1$   $7n^2 = n \times 7n$
   d) $10x = 5x \times 2$   $5x^2 = 5x \times x$
   e) $14y = 7y \times 2$   $7y^2 = 7y \times y$
   f) $4x^2 = 2x \times 2x$   $6x = 2x \times 3$

191

# Answers

**3** a) $4(t+2)$ b) $3(2-m)$ c) $2(1+9b)$
d) $5(x+2)$ e) $3(3z-11)$ f) $5(x-1)$
g) $2(a+3b+2c)$ h) $3(x-y+3z)$ i) $2(a-11b+3c)$
j) $2(4p+3p-2r)$ k) $7(2l-m-7n)$
l) $8(2a+3b-4c)$

**4** a) $x(2+x)$ b) $y(3-2y)$ c) $x(5x-4)$
d) $y(10y+7)$ e) $3(x^2+4x+2)$ f) $7(3+x-2x^2)$
g) $4(2g^2-4g-1)$ h) $5(3+x+4x^2)$
i) $6y(y-2x)$ j) $y^2(3x-11)$ k) $4t(s+2)$
l) $3a(b-2c+3d)$

**5** a) $4(2x+9y)$ b) $11(x+3)$ c) $5(3p+4q)$
d) $5(2x+y)$ e) $4(5a+4c)$ f) $5(l+n)$
g) $6(x+3y)$ h) $7(p+2q)$ i) $2(x-2y)$
j) $2(r+s)$ k) $x^2+9$ l) $x(x-8)$

## Page 121: Expanding two brackets

**1** a) $6a$ b) $6c^2$ c) $20y^2$
d) $-6x$ e) $-12x$ f) $15x$

**2** a) $x^2+4x+3$ b) $y^2+7y+10$ c) $12+7x+x^2$
d) $30+11y+y^2$ e) $x^2+x-2$ f) $y^2+y-6$
g) $x^2-4x-12$ h) $y^2-4y-5$ i) $x^2-9x+14$
j) $y^2-7y+12$ k) $x^2-8x+15$ l) $y^2-11y+30$

**3** a) $400+20+20+1=441$
b) $900+30+30+1=961$
c) $x^2+2x+1$ d) $1681$

**4** a) $a^2+6a+9$ b) $x^2+10x+25$ c) $a^2-6a+9$
d) $x^2-10x+25$ e) $4y^2+4y+1$ f) $9x^2+12x+4$

**5** a) $6x^2+19x+10$ b) $6x^2-19x+10$
c) $6x^2+11x-10$ d) $6x^2-11x-10$

**6** a) $xy+x+y+1$ b) $ad+2a+3d+6$
c) $xy-x-y+1$ d) $ck-10c+4k-40$
e) $ax+4a-3x-12$ f) $xy-x+10y-10$

**7** a) $(x+1)(x+2)$ b) $x^2, 2x, x, 2$
c) $x^2+2x+x+2 = x^2+3x+2$ d) C and B

## Page 123: Squares

**1** c) $4$ d) $4x^2$

**2** a) $9x^2$ b) $25y^2$ c) $9x^2$
d) $25y^2$ e) $16a^2$ f) $100u^2$

**3** a) $x^2+8x+16$ b) $y^2+6y+9$ c) $c^2-6c+9$
d) $n^2-10n+25$ e) $4x^2+20x+25$
f) $9+12t+4t^2$ g) $4y^2-12y+9$
h) $25d^2-10d+1$ i) $x^2-16$ j) $y^2-9$
k) $4x^2-25$ l) $9-4t^2$

**4** a) $x^2+2xy+y^2$ b) $x^2-2xy+y^2$ c) $x^2+14x+49$
d) $x^2-14x+49$ e) $4x^2+12x+9$
f) $16n^2-24n+9$ g) $25+10z+z^2$
h) $9-6x+x^2$ i) $4x^2+12xy+9y^2$
j) $4x^2-12xy+9y^2$ k) $x^2-y^2$ l) $x^2-49$
m) $4x^2-1$ n) $16n^2-81$ o) $25-z^2$
p) $9p^2-q^2$ q) $4x^2-9y^2$ r) $100a^2-4b^2$

**5** a) $(20-1)(20+1)$
b) $20^2-1^2 = 399$
c) $(30-1)(30+1) = 900-1$
          $= 899$

## Chapter 11: Probability

### Pages 126–127: Revision exercise

**1** A 0.5 (exact), B 0 (exact), C 1 (exact), D depends on time of year approx, E very small (approximate), F if marked should be at 1.

**2** a) $\frac{1}{3}$ b) $\frac{2}{3}$

**3** a) $\frac{100}{125} = \frac{4}{5}$ b) $\frac{20}{125} = \frac{4}{25}$ c) $\frac{4}{125}$ d) $\frac{1}{125}$

**4** a) $\frac{4}{30} = \frac{2}{15}$ b) $\frac{12}{30} = \frac{2}{5}$

c) 2 or 3 because the ratio on clear nights is 4 : 14 and the ratio on cloudy nights is probably similar.
d) No

**5** a)

red die

|  | 1 | 2 | 3 | 4 | 5 | 6 |
|---|---|---|---|---|---|---|
| 1 | 2 | 3 | 4 | 5 | 6 | 7 |
| 2 | 3 | 4 | 5 | 6 | 7 | 8 |
| 3 | 4 | 5 | 6 | 7 | 8 | 9 |
| 4 | 5 | 6 | 7 | 8 | 9 | 10 |
| 5 | 6 | 7 | 8 | 9 | 10 | 11 |
| 6 | 7 | 8 | 9 | 10 | 11 | 12 |

blue die

b) (i) $\frac{6}{36} = \frac{1}{6}$ (ii) $\frac{2}{36} = \frac{1}{18}$ (iii) $\frac{2}{36} = \frac{1}{18}$
(iv) $\frac{3}{36} = \frac{1}{12}$ (v) $\frac{6}{36} = \frac{1}{6}$

### Page 129: Two outcomes: 'either, or'

**1** a) $\frac{8}{52} = \frac{2}{13}$ b) $\frac{16}{52} = \frac{4}{13}$

**2** a) $\frac{5}{20} = \frac{1}{4}$ b) $\frac{15}{20} = \frac{3}{4}$ c) $\frac{13}{20}$ d) $\frac{7}{20}$

**3** a) $\frac{3}{65}$ b) $\frac{29}{65}$ c) $\frac{18}{65}$

**4** a) $\frac{6}{36} = \frac{1}{6}$ b) $\frac{15}{36} = \frac{5}{12}$ c) $\frac{9}{36} = \frac{1}{4}$ d) $\frac{15}{36} = \frac{5}{12}$

### Page 131: Two outcomes: 'first, then'

**1** a) $\frac{1}{4}$ or 0.25 b) $\frac{1}{2}$ or 0.5

**2** a) 0.1 b) 0.01 c) 0.009 d) 0.729

**3** a) $\frac{1}{8}$ b) $\frac{1}{8}$ c) $\frac{1}{8}$ d) $\frac{3}{8}$ e) $\frac{1}{8}$

**4** a) $\frac{1}{4}$ or 0.25 b) $\frac{1}{5}$ or 0.2 c) $\frac{1}{20}$ or 0.05
d) $\frac{1}{60}$ e) $\frac{3}{10}$

### Page 133: Probability trees

**1** a) 0.18 b) 0.28 c) 0.12 d) 0.42
**2** a) (i) 10 (ii) 30

# Answers

b) 

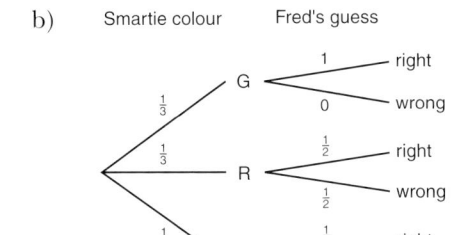

Smartie colour → Fred's guess

- G (1/3): right (1), wrong (0)
- R (1/3): right (1/2), wrong (1/2)
- B (1/3): right (1/2), wrong (1/2)

c) P(right) = $\frac{2}{3}$    d) Yes

## Chapter 12: Locus

### Page 137: Simple loci

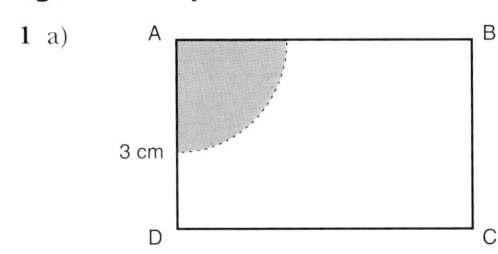

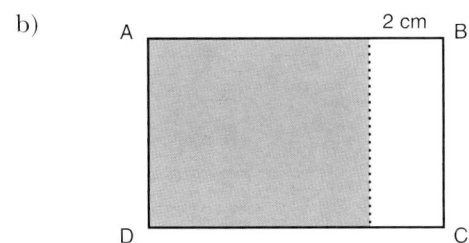

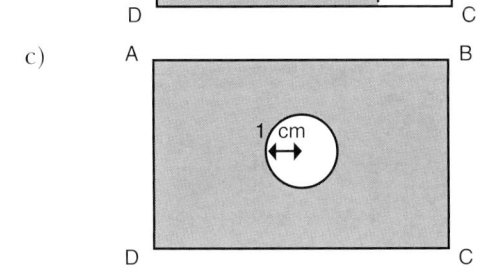

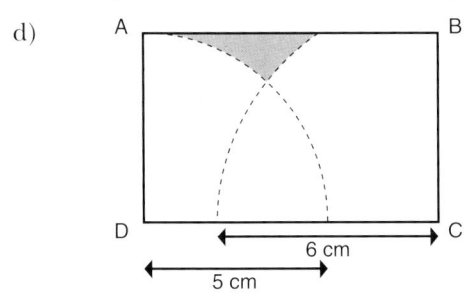

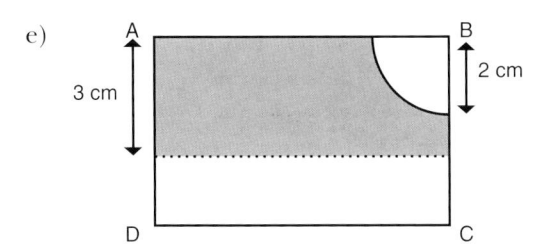

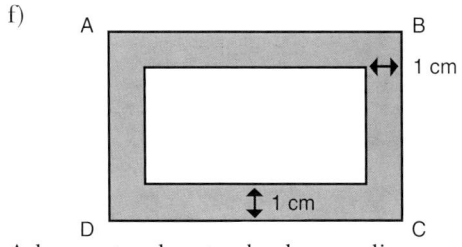

1 a) [diagram: rectangle ABCD with quarter-circle of radius 3 cm shaded at corner A]

b) [diagram: rectangle ABCD with region 2 cm from side BC shaded]

c) [diagram: rectangle ABCD with a circle of radius 1 cm unshaded in the middle]

d) [diagram: rectangle ABCD with two arcs from points 5 cm and 6 cm along DC intersecting]

e) [diagram: rectangle ABCD with region shaded 3 cm from top and 2 cm from B]

f) [diagram: rectangle ABCD with inner rectangle, 1 cm border]

2 Ask your teacher to check your diagram.
3 Ask your teacher to check your diagram.
4 Ask your teacher to check your diagram.
5 Ask your teacher to check your diagram.

### Page 139: A point equidistant from two fixed points

2 c) It is the centre of the circle.
3 Ask your teacher to check your diagram.
4 Ask your teacher to check your diagram.

### Page 141: A point equidistant from two lines

2 b) It is the same distance from all three sides. It is also the centre of a circle which just touches all three sides of the triangle.
3 a) b), c) d) Ask your teacher to check your diagrams.
4 Ask your teacher to check your diagram.

## Chapter 13: Quadratics

### Page 145: Factorising quadratic expressions

1 a) 1, 5
  b) $(x+1)(x+5)$
2 a) 3, 5
  b) $(y+3)(y+5)$
3 a) $(x+2)(x+7)$
  b) $(x+1)(x+14)$
  c) $(a+2)(a+5)$
  d) $(a+10)(a+1)$
  e) $(n+4)(n+1)$
  f) $(t+2)(t+2)$
  g) $(x+2)(x+2)$
  h) $(x+3)(x+6)$
  i) $(y+4)(y+6)$
4 a) −1, −5
  b) $(x-1)(x-5)$
5 a) −3, −5
  b) $(y-3)(y-5)$
6 a) $(x-1)(x-2)$
  b) $(x-3)(x-1)$
  c) $(x-5)(x-1)$
  d) $(x-11)(x-1)$
  e) $(x-7)(x-2)$
  f) $(x-2)(x-4)$
  g) $(p-3)(p-6)$
  h) $(a-4)(a-11)$
  i) $(x-44)(x-1)$

# Answers

**7** a) $(y-1)^2$
b) $(a+2)(a+11)$
c) $(x-3)(x-7)$
d) $(p-2)(p-8)$
e) $(t-2)(t-6)$
f) $(y+3)(y+4)$
g) $(a+1)^2$
h) $(x+3)^2$
i) $(y-2)^2$
j) $(t-5)^2$
k) $(x-3)^2$
l) $(y-1)(y-4)$
m) $(p-6)^2$
n) $(x+7)^2$
o) $(x-5)(x-20)$

## Page 147: More quadratic factorisation

**1** a) 6, 1
b) $(x+6)(x-1)$
c) $(x-6)(x+1)$

**2** a) 3, 5
b) $(y+5)(y-3)$
c) $(y-5)(y+3)$

**3** a) $(x+11)(x-1)$
b) $(x-11)(x+1)$
c) $(x+7)(x-1)$
d) $(x-7)(x+1)$
e) $(x+5)(x-1)$
f) $(x-5)(x+1)$
g) $(x+7)(x-2)$
h) $(x-7)(x+2)$
i) $(x+11)(x-8)$

**4** a) $(a+9)(a-2)$
b) $(a-9)(a+2)$
c) $(y+10)(y-1)$
d) $(y+2)(y-5)$
e) $(p-6)(p+3)$
f) $(x+3)(x-4)$
g) $(x+5)(x-4)$
h) $(a-2)(a+10)$
i) $(t+6)(t-2)$

**5** a) $(x-2)(x+2)$
b) $(y-5)(y+5)$
c) $(z-1)(z+1)$
d) $(n-4)(n+4)$
e) $(t-7)(t+7)$
f) $(p-10)(p+10)$

**6** a) $(x+1)(x+6)$
b) $(x+2)(x+4)$
c) $(r-4)(r-1)$
d) $(x-9)(x+1)$
e) $(y+4)(y-1)$
f) $(x+4)(x-3)$
g) $(t-4)(t+3)$
h) $(x-2)(x-9)$
i) $(p+6)(p-2)$
j) $(y-9)(y+9)$
k) $(b-5)(b+4)$
l) $(a-10)(a-1)$

**7** a) $4(x+4)(x-3)$
b) $3(a-1)(a+2)$
c) $3(x-2)(x+2)$
d) $3(x+1)^2$
e) $10(x-10)(x+10)$
f) $5(x-8)(x+10)$

## Page 149: Quadratic equations

**1** a) $x = 4$ or $1$
b) $x = 0$ or $4$
c) $x = 4$
d) $x = 0$ or $-4$
e) $y = 0$ or $3$
f) $x = 2$ or $3$
g) $x = 5$ or $-3$
h) $t = 0$ or $2$
i) $y = 0$ or $-4$

**2** a) 2, 7
b) 2, 5
c) 2, −7
d) −1, −4
e) 2, −3
f) 6, −2
g) 5, −10
h) 9, −7
i) 11, 1

**3** a) −5, 4
b) 2, −10
c) 1, 4
d) 4, −1
e) −2, −9
f) 4, −3
g) 2, −8
h) 2, −2
i) 5, 6

**4** a) (i) $w^2 + 3w - 54 = 0$ (ii) $w = 6$
b) (i) $8l^2 - 80l - 3000 = 0$ (ii) $l = 25$
c) (i) $2x^2 + 8x - 90 = 0$ (ii) $x = 5$

**5** a)

| $x$ | −4 | −3 | −2 | −1 | 0 | 1 | 2 | 3 | 4 |
|---|---|---|---|---|---|---|---|---|---|
| $x^2$ | 20 | 12 | 6 | 2 | 0 | 0 | 2 | 6 | 12 |

b) −2, 3
c) −2, 3
d) $x^2 - x = 6$ when $y = 6$ on the graph.
e) 3.4, −2.4
It will not factorise

# Answers

## Chapter 14: Transformations

### Page 153: Revision exercise

1 Ask your teacher to check your drawings
2 Ask your teacher to check your drawings
3 a) C, D, G, I   b) E, F, H
   c) (i) Translation $\begin{pmatrix}1\\5\end{pmatrix}$
       (ii) Reflection in $x = -3$
       (iii) Rotation, centre $(-3, -3)$ 90° clockwise
4 Ask your teacher to check your graph.
   e) An octagon. (Note it is not quite regular)
      Mirror Symmetry in $x$ axis, $y$ axis, $y = x$ and $y = -x$
      Rotational symmetry about origin, order 4

### Page 155: Translations using column vectors

1 a) $\begin{pmatrix}5\\-1\end{pmatrix}$   b) $\begin{pmatrix}4\\2\end{pmatrix}$   c) $\begin{pmatrix}-5\\-3\end{pmatrix}$   d) $\begin{pmatrix}-2\\3\end{pmatrix}$
   e) $\begin{pmatrix}0\\4\end{pmatrix}$   f) $\begin{pmatrix}-2\\-5\end{pmatrix}$   g) $\begin{pmatrix}3\\-6\end{pmatrix}$   h) $\begin{pmatrix}-7\\0\end{pmatrix}$
2 a) Ask your teacher to check your drawings
   b) $\begin{pmatrix}5\\-3\end{pmatrix}$
   c) Ask your teacher to check your drawings
   d) $\begin{pmatrix}2\\4\end{pmatrix}$   e) $\begin{pmatrix}-7\\-1\end{pmatrix}$   f) $\begin{pmatrix}7\\1\end{pmatrix}$

### Page 157: Reflection

1 Ask your teacher to check your drawings
2 Ask your teacher to check your drawings
3 a) (1, 3)   (−1, 4)   (−2, 2)
   c) (3, 1)   (4, −1)   (2, −2)
   e) (−3, −1) (−4, 1)   (−2, 2)

### Page 159: Rotation

1 Ask your teacher to check your drawings
2 a) Reflection in $y = x$
   b) Rotation through 90° clockwise about (2, −1)
   c) Reflection in $x = \frac{1}{2}$
   d) Translation $\begin{pmatrix}-1\\7\end{pmatrix}$
   e) Rotation through 180° about (−5, 0)
   f) Reflection in $y = -x$
   g) Translation $\begin{pmatrix}-5\\-6\end{pmatrix}$
   h) Reflection in $y = 2$
   i) Rotation through 90° clockwise about (0, 3)

### Page 161: Combining transformations

1 a) (i) Reflection in $y$ axis
       (ii) Reflection in $x$ axis
       (iii) Rotation centre (0, 0) 180°
   b) (i) Reflection in $y$ axis
       (ii) Reflection in $x = 1.5$
       (iii) Translation $\begin{pmatrix}3\\0\end{pmatrix}$
   c) (i) Yes   (ii) Yes   (iii) No
2 a) (i) Translation $\begin{pmatrix}3\\1\end{pmatrix}$
       (ii) Translation $\begin{pmatrix}1\\3\end{pmatrix}$
       (iii) Translation $\begin{pmatrix}4\\4\end{pmatrix}$
   b) (i) Yes   (ii) No   (iii) No
3 a) (i) Enlargement, centre (3, −2), scale factor 2
       (ii) Rotation, centre (−1, 4), 90° anticlockwise
   b) (i) Rotation, centre (0, 0), 90° anticlockwise
       (ii) Enlargement, centre (−1, −2), scale factor 2
   c) No
4 (i) A → C → E   (ii) A → H → A

## Chapter 15: Fractions in algebra

### Pages 164–165: Reminder

1 a) 8   b) 9   c) 12   d) 9   e) 3
   f) 3   g) 7   h) 5
2 a) 16  b) 12  c) 15   d) 15  e) 7
   f) 6   g) 4   h) 7
3 a) 16  b) 12  c) 15   d) 15  e) 7
   f) 6   g) 4   h) 7
4 a) 8   b) 9   c) 12   d) 9   e) 3
   f) 3   g) 4   h) 5
5 a) (i) true    (ii) false   (iii) true
       (iv) false   (v) false    (vi) true
   b) (i) $4x$   (ii) $x$    (iii) $7x$   (iv) $35x$
       (v) $33x$  (vi) $3x$   (vii) $x$    (viii) $3x$
6 a) $a = 28$   b) $b = 12$   c) $c = 16$   d) $d = 9$
   e) $e = 25$   f) $f = 49$
7 a) $a = 24$   b) $b = 6$    c) $c = 15$   d) $d = 18$
   e) $e = 45$   f) $f = -4$
8 a) $x = 5$    b) $x = 19$   c) $x = 8$    d) $x = 25$
   e) $x = 12$   f) $x = 6$
9 a) $x = 4$    b) $x = 2$    c) $x = 1$    d) $x = 17$
   e) $x = 8$    f) $x = 9$

### Page 167: Indices

1 a) (i) $5^4$ (ii) $7^6$   (iii) $3^5$
   b) (i) $a^4$ (ii) $b^6$   (iii) $c^5$

# Answers

2  a) $9 \times 9 \times 9$
   b) $7 \times 7 \times 7 \times 7$
   c) $5 \times 5 \times 5 \times 5 \times 5 \times 5 \times 5 \times 5$
   d) $4 \times 4$
   e) $3 \times 3 \times 3 \times 3 \times 3$
   f) $6 \times 6 \times 6 \times 6 \times 6 \times 6 \times 6$
   g) $a \times a \times a$
   h) $b \times b \times b \times b$
   i) $c \times c \times c \times c \times c \times c \times c \times c$
   j) $3 \times d \times d$
   k) $7 \times x \times x \times x \times x \times x$
   l) $4 \times y \times y \times y \times y \times y \times y \times y$

3  a) $3^7$  b) $5^7$  c) $4^2$  d) $3$  e) $x^7$  f) $y^8$
   g) $a^4$  h) $b$

4  a) $8x$  b) $21y$  c) $15x$  d) $18x^2$  e) $12x^2$  f) $12y^3$
   g) $30z^4$  h) $16x^3$

5  a) $a^2$  b) $b$  c) $c^4$  d) $d^2$  e) $x^5$  f) $x^4$
   g) $x^5$  h) $y^5$

6  a) $3a^2$  b) $4b$  c) $3c^4$  d) $3d^2$  e) $\frac{1}{4}x^5$  f) $\frac{1}{5}x^4$
   g) $\frac{1}{3}x^5$  h) $\frac{1}{4}y^5$

7  a) $x$  b) $x^2$  c) $a^5$  d) $a^2$  e) $\frac{6b}{a}$  f) $\frac{2cd}{5}$
   g) $\frac{7ab^2}{5c}$  h) $4r$

## Page 169: Rational functions

1  a) $x$  b) $x$  c) $x^2$  d) $x^2$
   e) $x$  f) $\frac{1}{x}$

2  a) (i) $x(3x+1)$  (ii) $x^2(4x+1)$  (iii) $x^2(5+x)$
   b) (i) $3x+1$  (ii) $4x+1$  (iii) $5+x$

3  a) $x-5$  b) $x-2$  c) $x+3$
   d) $x+1$  e) $x-2$  f) $x-2$

4  a) $\frac{x-5}{x+5}$  b) $\frac{x-2}{x+2}$  c) $\frac{x+3}{x-3}$

5  a) (i) $x(x+3)$  (ii) $(x-1)(x+1)$  (iii) $x(x-9)$
      (iv) $x(3x+2)$  (v) $x^2(x+1)$
   b) (i) $x$  (ii) $x-1$  (iii) $x$
      (iv) $\frac{x}{3x+2}$  (v) $\frac{x}{x+1}$

6  a) (i) $(x+2)(x+5)$  (ii) $(x-2)(x-3)$
      (iii) $(x+3)(x-1)$
   b) (i) $x+5$  (ii) $x-2$  (iii) $x+3$

## Chapter 16: Enlargement and similarity

### Pages 172–173: Revision exercise

1  B - scale factor 3, C - scale factor 2
   A - area 10, perimeter 14
   B - area 90, perimeter 42
   C - area 40, perimeter 28
2  Ask your teacher to check your drawings.

### Page 175: Centres of enlargement

1  Ask your teacher to check your drawings.
2  Ask your teacher to check your drawings.
3  a) Scale factor 2, centre (1, 8)
   b) Scale factor 4, centre (4, 8)

### Page 177: Scale factors less than 1

1  a) 54 cm  b) 4.5 cm  c) 27 cm  d) 12 cm
2  Ask your teacher to check your diagram.
3  a) 2  b) $\frac{1}{3}$  c) 3  d) $\frac{1}{2}$
4  a) Scale factor $\frac{1}{3}$, centre (0, 1)
   b) Scale factor $\frac{1}{2}$, centre (3, 4)

### Page 179: Similar shapes

1  a) 4/3 or 1.33  b) 2.25  c) 0.375
2  a) B, D, E  b) 30 cm
3  a) A and F, B and D, C and E.

### Page 181: Using similarity

1  a) $x = 4.5$ cm, $y = 6$ cm
   b) $a = 2$ cm, $b = 7.5$ cm
   c) $p = 3.2$ cm, $q = 2$ cm
   d) $d = 4.8$ cm, $e = 3.33$ cm, $f = 3.75$ cm
   e) $x = 2.5$ cm, $y = 10.5$ cm
   f) $p = 5.33$ cm, $q = 6.67$ cm
2  9.075 m
3  A0   1188   840
   A1   840    594
   A2   594    420
   A3   420    297
   A4   297    210
   A5   210    148
   A6   148    105

# Index

## A
adjacent sides 70
algebra fractions 164–169
   rational function 168–169
angle bisector 140–141
angles
   in a circle 110–111
   in a polygon 44–47
   in a triangle 38–39

## B
BIDMAS 20
bisector
   angle 140–141
   perpendicular 138–139
brackets
   expanding 120–121
   squaring 122–123

## C
circles 108–113
   angles in 110–111
   cyclic quadrilateral 108
   segments 110
   shapes in 108–109
   tangents 112–113
column vectors 154–155
congruent shapes 152
co-ordinates 50–53
cosine(cos) 78–79, 80–81
cyclic quadrilateral 108

## D
dividing fractions 8–9

## E
'either, or' 128–129
enlargement 172–181
   centres of 174–175
   scale factors 176–177
   similar shapes 178–181
equations 20–33
   formulae 20–23, 32–33
   graphs 28–31, 66
   multiplying 64–65
   problem solving 26–27
   quadratic 148–149
   simultaneous 60–67
   substitution 66-67
   trial and improvement 30
expanding two brackets 120–121
expressions, manipulating 116–123
   difference of two squares 122
   expanding two brackets 120–121
   factorising 118–119
   FOIL 120
   like terms 116–117
   squaring brackets 122–123
   unlike terms 116–117

## F
factorising 118–119, 144–147
'first, then' 130–131
FOIL 120
formulae 20–23, 32–33
   making up 22–23
   rearranging 32–33
   subject of 22
fractions 4-9, 164–169
   dividing 8–9
   improper 4
   in algebra 164–169
   multiplying 6–7

## G
graphs 50–57, 92–95
   equation solving 28–31, 66
   inequalities 92–95
   sequences 56–57
   using co-ordinates 50–53

## H
hypotenuse 70

## I
improper fractions 4
index form 102–103
indices 100–105, 166–167
   index form 102–103
   rules 102
   standard form 103–105
inequalities 86–97
   graphs 92–95
   number lines 88–89
   solution sets 96–97
   solving 90–91
   symbols 86–87

## L
like terms 116–117
locus 136–141
   angle bisector 140–141
   perpendicular bisector 138–139

## M
manipulating expressions 116–123
mixed numbers 4, 6–7
multiplying
   equations 64–65
   fractions 6–7

## N
number lines, inequalities 88–89

## O
opposite sides 70
original price 14-15

## P
percentages 10–16
perpendicular bisector 138–139
polygons 44–47
   angles of 44–47
   exterior angles of 44
   interior angles of 44
   irregular 44
   regular 44
probability 126–133
   'either, or' 128–129
   'first, then' 130–131
   tree diagrams 132–133

## Q
quadratics 144–149
   equation 148–149
   factorising 144–147

## R
rational function 168–169
reflection 152, 156–157, 158–159
rotation 152, 158–159

# Index

## S
scale factors 176–177
segments of a circle 110
sequences 56–57
similarity 178–181
similar triangles 70
simultaneous equations 60–67
sine(sin) 76–77, 80–81
SOHCAHMA 80
solution sets 96–97
solving inequalities 90–91
squaring brackets 122–123
standard form, indices 103–105
subject of a formula 22
substitution in equations 66–67
symbols, inequalities 86–87

## T
tangent(tan) 72–75, 80–81
tangents 112–113
transformations 152–161
   column vectors 154–155
   combining 160–161
   congruent shapes 152
   reflection 152, 156–157, 158–159
   rotation 152, 158-159
   translation 152, 154–155, 158–159
translation 152, 154–155, 158–159
tree diagrams, probability 132–133
trial and improvement 30
triangles 36–43
   drawing 40–43
   hypotenuse 70
   interior angles in 36–39
   opposite angles in 36–39
   similar 70
trigonometry 70–83
   cosine(cos) 78–81
   sine(sin) 76–77, 80–81
   SOHCAHTOA 80
   tangent(tan) 72–75, 80–81

## U
unknowns 24–25
unlike terms 116–117